Combinatorics
on
Words

GIAN-CARLO ROTA, *Editor*
ENCYCLOPEDIA OF MATHEMATICS AND ITS APPLICATIONS

GIAN-CARLO ROTA, *Editor*
ENCYCLOPEDIA OF MATHEMATICS AND ITS APPLICATIONS

GIAN-CARLO ROTA, *Editor*

ENCYCLOPEDIA OF MATHEMATICS AND ITS APPLICATIONS

ENCYCLOPEDIA OF MATHEMATICS and Its Applications

GIAN-CARLO ROTA, Editor
Department of Mathematics
Massachusetts Institute of Technology
Cambridge, Massachusetts

Editorial Board

GIAN-CARLO ROTA, *Editor*
ENCYCLOPEDIA OF MATHEMATICS AND ITS APPLICATIONS
Volume 17

Section: Algebra
P. M. Cohn and Roger Lyndon, *Section Editors*

Combinatorics on Words

M. Lothaire

Foreword by
Roger Lyndon
University of Michigan

1983

Addison-Wesley Publishing Company
Advanced Book Program/World Science Division
Reading, Massachusetts

London · Amsterdam · Don Mills, Ontario · Sydney · Tokyo

Library of Congress Cataloging in Publication Data

Lothaire, M.
   Combinatorics on words.

   (Encyclopedia of mathematics and its applications;
v. 17)
   Bibliography: p.
   Includes index.
   1. Combinatorial analysis. 2. Word problems
(Mathematics) I. Title. II. Series.
   QA164.L67   511′.6   82-6716
   ISBN 0-201-13516-7   AACR2

American Mathematical Society (MOS) Subject Classification Scheme (1980): 05-01, 05A99,
20MO5, 68-01, 68E99.

ABCDEFGHIJ—HA—898765432

Manufactured in the United States of America

Contents

Editor's Statement

A large body of mathematics consists of facts that can be presented and described much like any other natural phenomenon. These facts, at times explicitly brought out as theorems, at other times concealed within a proof, make up most of the applications of mathematics, and are the most likely to survive change of style and of interest.

This ENCYCLOPEDIA will attempt to present the factual body of all mathematics. Clarity of exposition, accessibility to the non-specialist, and a thorough bibliography are required of each author. Volumes will appear in no particular order, but will be organized into sections, each one comprising a recognizable branch of present-day mathematics. Numbers of volumes and sections will be reconsidered as times and needs change.

It is hoped that this enterprise will make mathematics more widely used where it is needed, and more accessible in fields in which it can be applied but where it has not yet penetrated because of insufficient information.

GIAN-CARLO ROTA

Section Editor's Foreword

This is the first book devoted to broad study of the combinatorics of *words*, that is to say, of sequences of symbols called *letters*. This subject is in fact very ancient and has cropped up repeatedly in a wide variety of contexts. Even in the most elegant parts of abstract pure mathematics, the proof of a beautiful theorem surprisingly often reduces to some very down to earth combinatorial lemma concerning linear arrays of symbols. In applied mathematics, that is in the subjects to which mathematics can be applied, such problems are even more to be expected. This is true especially in those areas of contemporary applied mathematics that deal with the discrete and non-commutative aspects of the world about us, notably the theory of automata, information theory, and formal linguistics.

The systematic study of words seems to have been initiated by Axel Thue in three papers [Norske Vid. Selsk. Skr. I Mat. Nat. Kl., Christiania, 1906, 1–22; 1912, 1–67; 1914, 1–34.]. Even more than for his theorems, we owe him a great debt for delineating this subject. Both before and after his time, a multitude of fragmentary results have accumulated in the most diverse contexts, and a substantial but not very widely known lore was beginning to crystallize to the point where a systematic treatment of the subject was badly needed and long over due.

This need is splendidly fulfilled by the present volume. It provides a clear and easily accessible introduction to the area, treating in some depth a representative selection of the most important branches of the subject. In particular, connections with free Lie algebras and algebras with polynomial identities are treated in full. The Preface by Dominique Perrin gives a lucid account of this book, and we need not say more on that matter. However, we want to amplify his remarks on the origins of this undertaking.

First, Marcel P. Schützenberger should be acknowledged as the "grandfather" of the book. It was Marco who initiated the systematic combinatorial and algebraic study of monoids, the natural habitat of words, and of the connections and applications of this subject to such classical areas as group representation theory, infinite groups, Lie algebras, probability theory, as well as to more recent "applied" subjects, notably computer science and mathematical linguistics. In addition to his own important and seminal work in these subjects, it was he who founded the most important school dealing with these and related subjects (now the Laboratoire Informatique Théorique et Programmation at Paris). Most of the contributors to this book are his former students, or students of theirs, and all are disciples of his teachings. Today, when he is not promulgating the virtues of monoids in hazardous foreign climes, he is walking the corridors of Paris VII stimulating the workers there and instigating new lines of research.

Second, this book is the result of a friendly collaboration of the group of authors that have realized it. This collective enterprise was initiated by Jean François Perrot and led to its conclusion by Dominique Perrin, who played the role of editor of the volume. He can be considered to be the "biological father" of this book, and I have been privileged to see him proving theorems, testing conjectures, as well as phoning long distance to obtain copy, editing copy, and carrying it to the post office. It is to this team that we are indebted for both the existence and the high quality of the present work.

It is a pleasure to witness such an auspicious official inauguration of a newly recognized mathematical subject, one which carries with it certain promise of continued increasingly broad development and application.

ROGER LYNDON

Preface

Combinatorics on words is a field that has grown separately within several branches of mathematics, such as group theory or probabilities, and appears frequently in problems of computer science dealing with automata and formal languages. It may now be considered as an independent theory because of both the number of results that it contains and the variety of possible applications.

This book is the first attempt to present a unified treatment of the theory of combinatorics on words. It covers the main results and methods in an elementary presentation and can be used as a textbook in mathematics or computer science at undergraduate or graduate level. It will also help researchers in these fields by putting together a lot of results scattered in the literature.

The idea of writing this book arose a few years ago among the group of people who have collectively realized it. The starting point was a mimeographed text of lectures given by M. P. Schützenberger at the University of Paris in 1966 and written down by J. F. Perrot. The title of this text was "Quelques Problèmes combinatoires de la théorie des automates." It was widely circulated and served many people (including most of the authors of this book) as an introduction to this field. It was J. F. Perrot's idea to make a book out of these notes, whose title varied from *The Little Red Book* in the sixties (by the color of its cover, later selected for this series) to *The Apocryphal*, a name that could dilute the responsibility for mistakes, if any, in the text.

Let us put aside for now the domestic history of this book and turn to its subject. The objects considered by people who study combinatorics on words are words, that is to say, sequences of elements taken from a set. Typical phenomena that can be observed in a word are certain kinds of repetitions, decompositions into words of a special sort, and the results of rearrangement of the letters. The type of results obtained is perhaps reminiscent of the beginnings of number theory.

The first significant works on the subject go back to the start of this century, appearing in A. Thue's papers on square-free words and MacMahon's treatise on combinatory analysis. Apart from pure combinatorics this kind of problem has also been studied by scholars dedicated to probabilities, especially to fluctuations of random variables. In pure algebra, problems on words appear in a number of situations, including free algebras, free groups, and free semigroups. More recently, the theory of automata and formal languages was developed, taking inspiration from problems of computer science. For these problems, words are the natural object, because any computation gives rise to a sequence of elementary operations and also because a sequential machine can process an object only if it has a linear structure, which again is a word. The same observations obviously apply to natural languages. In fact, the beginnings of automata theory and of modern linguistics are interconnected.

The main originality of this book is that it gathers for the first time the pieces of the jigsaw puzzle just described. It is dedicated neither to group or semigroup theory nor to automata theory but only to words (although the origin or the consequences of the methods and results presented are explicitly mentioned all along). The subtlest difference between the subject of this book and any of the aforementioned theories is perhaps with automata theory. It can be roughly said that automata theory (and formal language theory) deals with sets of words whereas combinatorics on words considers properties of *one* word. This distinction is sometimes rather artificial, however, and the situation is the same as for determining what the work "combinatorial" exactly means. We have put aside another subject that is rather wide and also closely related to this one, the theory of codes, which will be the subject of another book. Let me now briefly present the contents of this book.

Chapter 1 contains the main definitions, together with some elementary properties of words frequently used in the sequel. The following three chapters deal with *unavoidable regularities*, which are properties of words that become true when their length tends to infinity. It is therefore a study of asymptotic properties of words.

Chapters 5–7 may be considered another block. They deal with properties of words related with classical noncommutative algebra. The first, Chapter 5, treats Lie algebras, the second is linked with nilpotent groups, and the third algebras with polynomial identities. The rest of the book, Chapters 8–11, deals with specific aspects of words, each worthy of a complete volume.

This book is written at a completely elementary level so as to be accessible to anyone with a standard mathematical background. The authorship of each chapter is different, but the notation is uniform throughout and the architecture of the book (including use of results from one chapter in another) is the result of the joint conception of the coauthors.

Each chapter ends with a series of problems. Either they comment upon particular cases of the results of the chapter, or, more often, they mention some additional results. Difficult ones are indicated with an asterisk or double asterisk.

It is a pleasure to express the thanks of the authors of this book for the collaboration that we received during its preparation. Dorothée Reutenauer translated the paper of Shirshov on which Chapter 7 relies. Howard

Straubing read several chapters and formulated useful comments. Volker Strehl carefully read Chapter 10 and helped with suggestions. We are also indebted to Georges Hansel, Gérard Lallement, André Lentin, Roger Lyndon, and Maurice Nivat and, of course, to Jean-François Perrot whose work was the starting point for this enterprise. The help of Claudine Beaujean, Maryse Brochot, Martine Chaumard, Monique Claverie, Arlette Dupont and Sylvie Lutzing, in particular for typing, is gratefully acknowledged.

To close this preface, I should like to mention the belief my coauthors and I share that this book will serve as an incentive for further developments of this beautiful theory. It might be the case that a good part of it will be superseded within a few years, and this is exactly what we hope.

DOMINIQUE PERRIN

Combinatorics
on
Words

_____

Words

1.0. Introduction

This chapter contains the main definitions used in the rest of the book. It also presents some basic results about words that are of constant use in the sequel. In the first section are defined words, free monoids, and some terms about words, such as length and factors.

Section 1.2 is devoted to submonoids and to morphism of free monoids, one of the basic tools for words. Many of the proofs of properties of words involve a substitution from the alphabet into words over another alphabet, which is just the definition of a morphism of free monoids. A nontrivial result called the *defect theorem* is proved. The theorem asserts that if a relation exists among words in a set, those words can be written on a smaller alphabet. This is a weak counterpart for free monoids of the Nielsen–Schreier theorem for subgroups of a free group.

In Section 1.3 the definition of conjugate words is given, together with some equivalent characterizations. Also defined are *primitive words*, or words that are not a repetition of another word. A very useful result, due to Fine and Wilf, is proved that concerns the possibility of multiple repetitions. The last section introduces the notation of formal series that deal with linear combinations of words, which will be used in Chapters 5–7 and 11.

A list of problems, some of them difficult, is collected at the end. Two of them (1.1.2 and 1.2.1) deal with free groups; their object is to point out the existence of a combinatorial theory of words in free groups, although the theory is not developed in the present book (see Lyndon and Schupp 1977). Two others (1.1.3 and 1.3.5) deal with the analysis of algorithms on words.

1.1. Free Monoids and Words

Let A be a set that we shall call an *alphabet*. Its elements will be called *letters*. (In the development of this book, it will often be necessary to suppose that the alphabet A is finite. Because this assumption is not always necessary, however, it will be mentioned explicitly whenever it is used.)

A word over the alphabet A is a finite sequence of elements of A:

$$(a_1, a_2, \ldots, a_n), \quad a_i \in A.$$

The set of all words over alphabet A is denoted by A^*. It is equipped with a binary operation obtained by concatenating two sequences.

$$(a_1, a_2, \ldots, a_n)(b_1, b_2, \ldots, b_m) = (a_1, a_2, \ldots, a_n, b_1, b_2, \ldots, b_m).$$

This binary operation is obviously associative, which allows writing a word as

$$a_1 a_2 \cdots a_n$$

instead of

$$(a_1, a_2, \ldots, a_n),$$

by identifying a letter $a \in A$ with the sequence (a).

The empty sequence, called the *empty word*, is a neutral element for the operation of concatenation. It is denoted by 1; hence, for any word w

$$1w = w1 = w.$$

A *monoid* is a set M with a binary operation that is associative and has a neutral element denoted by 1_M. Hence, what has been defined on the set A^* is a monoid structure.

A *morphism* of a monoid M into a monoid N is a mapping φ of M into N compatible with operations of M and N:

$$\varphi(mm') = \varphi(m)\varphi(m'), \quad m, m' \in M,$$

and such that $\varphi(1_M) = 1_N$.

PROPOSITION 1.1.1. *For any mapping α of A into a monoid M, there exists a unique morphism φ of monoids from A^* into M such that the following diagram is commutative:*

where i is the natural injection of A into A^*.

Proof. Left to the reader.   ∎

Because of this property (called a *universal property*), the set A^* of all words over the alphabet A is called the *free monoid* over the set A.

The set of all nonempty words over A will be denoted by A^+:

$$A^+ = A^* - 1.$$

It is called the *free semigroup* over A (recall that a semigroup is a set with an associative binary operation). It may be readily verified that Proposition 1.1.1 can be stated for A^+ instead of A^* by replacing the term "monoids" by "semigroups."

As for any monoid the binary operation of A^* may by extended to the subsets of A^* by defining for X, $Y \subset A^*$.

$$XY = \{xy \mid x \in X, y \in Y\}.$$

We shall come back to this extension in Section 1.4. Consider now some terminology about words.

The *length* of the word $w = a_1 a_2 \cdots a_n$, $a_i \in A$ is the number n of the letters w is a product of. It will be denoted by $|w|$:

$$|w| = n.$$

The length of the empty word is 0 and the mapping $w \mapsto |w|$ is a morphism of the free monoid A^* onto the additive monoid $\mathbb{N}$ of positive integers.

For a subset B of the alphabet A, we denote by $|w|_B$ the number of letters of w that belong to B. Therefore,

$$|w| = \sum_{a \in A} |w|_a.$$

Denoted by $\mathrm{alph}(w)$ is the subset of the alphabet formed by the letters actually occurring in w. Therefore $a \in A$ belongs to $\mathrm{alph}(w)$ iff

$$|w|_a \geq 1.$$

A word $v \in A^*$ is said to be a *factor* of a word $x \in A^*$ if there exist words $u, w \in A^*$ such that

$$x = uvw.$$

The relation "v is factor of x" is an order on A^*. A factor v of $x \in A^*$ is said to be *proper* if $v \neq x$:

A word v is said to be a *left factor* of $x \in A^*$ if there exists a word $w \in A^*$ such that

$$x = vw,$$

and it is said to be a *proper left factor* if $v \neq x$. The relation "v is a left factor of x" is again an order on A^*; it will be denoted by

$$v \leqslant x.$$

This order has the fundamental property that if

$$v \leqslant x, \qquad v' \leqslant x,$$

then v and v' are comparable: $v \leqslant v'$ or $v' \leqslant v$.

More precisely, if

$$vw = v'w',$$

either there exists $s \in A^*$ such that $v = v's$ (and then $sw = w'$) or there exists $t \in A^*$ such that $v' = vt$ (and then $w = tw'$). This will be referred to as the property of *equidivisibility* of the free monoid.

The definition of a *right factor* is symmetrical to that of a left factor. The *reversal* of a word $w = a_1 a_2 \cdots a_n$, $a_i \in A$, is the word

$$\tilde{w} = a_n \cdots a_2 a_1.$$

Hence v is a left factor of x iff $\tilde{v}$ is a right factor of $\tilde{x}$. We shall also use the notation $w^{\sim}$ instead of $\tilde{w}$; we may then write or all $u, v \in A^+$,

$$(uv)^{\sim} = \tilde{v}\tilde{u}.$$

A word w is palindrome if $w = \tilde{w}$.

A word $v \in A^*$ is said to be a *subword* of a word $x \in A^*$ if

$$v = a_1 a_2 \cdots a_n, \qquad a_i \in A, n \geqslant 0,$$

and there exist $y_0, y_1, \ldots, y_n \in A^*$ such that

$$x = y_0 a_1 y_1 a_2 \cdots a_n y_n.$$

Therefore v is a subword of x if it is a sub-sequence of x.

1.2. Submonoids and Morphisms

A submonoid of a monoid M is a subset N of M containing the neutral element of M and closed under the operation of M: $NN \subset N$. Given a sub set X of the free monoid A^*, we denote by X^* the submonoid of A^* generated by X. Conversely, given a submonoid P of A^*, there exists a unique set X that generates P and is minimal for set-inclusion. In fact, X is the set

$$X = (P - 1) - (P - 1)^2$$

of the nonempty words of P that cannot be written as the product of two nonempty words of P. It is a straightforward verification that X generates P and that it is contained in any set $Y \subset A^*$ generating P. The set X will be referred to as the *minimal generating set* of P.

A monoid M is said to be *free* if there exist an alphabet B and an isomorphism of the free monoid B^* onto M. For instance, for any word $w \in A^+$ the submonoid generated by w, written w^* instead of $\{w\}^*$, is free. It is very important to observe that not all the submonoids of a free monoid are themselves free (see Example 1.2.2).

PROPOSITION 1.2.1. *Let P be a submonoid of A^* and X be its minimal generating set. Then P is free iff any equality*

$$x_1 x_2 \cdots x_n = y_1 y_2 \cdots y_m, \qquad n, m \geq 0, \quad x_i, y_j \in X$$

implies $n = m$ and $x_i = y_i$, $1 \leq i \leq n$.

The proof is again left to the reader. The minimal generating set of a free submonoid P of A^* is called a *code*; it is referred to as the *basis* of P.

A set $X \subset A^*$ is called a *prefix* if for $x, y \in X$.

$$x \leq y$$

implies $x = y$; it can easily be verified that any prefix $X \subset A^+$ is a code.

Example 1.2.2. Let $A = \{a, b\}$; the set $X = \{a, b, ab\}$ is not a code since it is not the minimal generating set of X^*. The set $Y = \{a, ab, ba\}$ is the minimal generating set of Y^*; yet it is not a code because

$$a(ba) = (ab)a$$

is a nontrivial equality between products of elements of Y. The set $Z = \{aa, ba, baa, bb, bba\}$ can be verified to be a code.

The following characterization of free submonoids of A^* is useful:

PROPOSITION 1.2.3. *A submonoid P of A^* is free iff for any word $w \in A^*$, one has $w \in P$ whenever there exist $p, q \in P$ such that*

$$pw, wq \in P.$$

Proof. Let P be a submonoid of A^* and denote by X its minimal generating set. First suppose that the preceding condition holds for P. Then if

$$x_1 x_2 \cdots x_n = y_1 y_2 \cdots y_m, \quad x_i \in X, \quad y_j \in X, \tag{1.2.1}$$

we may suppose that $x_1 \leqslant y_1$ and let $y_1 = x_1 w$, $w \in A^*$. Then

$$x_2 \cdots x_n = w y_2 \cdots y_n,$$

and therefore $x_1 w, w y_2 \cdots y_n \in P$; this implies by hypothesis $w \in P$. Since X is the minimal generating set of P, we have $w = 1$, and this proves that eq. (1.2.1) is trivial by induction on $n + m$. Therefore P is free.

Conversely, if P is free, let φ by an isomorphism of a free monoid B^* onto P, with $X = \varphi(B)$. Then if for $p, q \in P$, one has $pw, wq \in P$, let $\varphi(x) = p$, $\varphi(y) = wq$, $\varphi(z) = pw$, $\varphi(t) = q$. Since $\varphi(xy) = \varphi(zt)$ we have $xy = zt$, and this implies that $z = xu$, $u \in B^*$. Therefore $w = \varphi(u) \in P$. ∎

COROLLARY 1.2.4. *An intersection of free submonoids of A^* is free.*

Proof. If the submonoids P_i, $i \in I$ are free, and if there exists

$$p, q \in P = \bigcap_{i \in I} P_i,$$

such that $pw, wq \in P$, then by Proposition 1.2.3, $w \in P_i$ for each $i \in I$ and therefore $w \in P$. By 1.2.2, this shows that P is free. ∎

If X is any subset of A^*, the set $\mathcal{F}$ of free submonoids of A^* containing X is not empty (it contains A^*) and, by Corollary 1.2.4, it is closed under intersection. Therefore the intersection of all elements of $\mathcal{F}$ is the smallest free submonoid containing X; the code generating this submonoid is called the *free hull* of X.

THEOREM 1.2.5 (*Defect theorem*). *The free hull Y of a finite subset $X \subset A^*$, which is not a code, satisfies the inequality*

$$\mathrm{Card}(Y) \leqslant \mathrm{Card}(X) - 1.$$

Proof. Consider the mapping α of X into Y associating to $x \in X$ the word $y \in Y$ such that $x \in y Y^*$; since Y is a code, the mapping α is well defined.

As X is not a code, there exists an equality

$$x_1 x_2 \cdots x_n = y_1 y_2 \cdots y_n$$

with $x_i, y_j \in X$ and $x_1 \neq y_1$. Therefore, $\alpha(x_1) = \alpha(y_1)$ and α cannot be injective.

The following shows that α is surjective: if it were not, let $z \in Y$ be such that $z \notin \alpha(X)$; consider the set

$$Z = (Y - z) z^*.$$

The set Z is a code since an equality

$$z_1 z_2 \cdots z_n = z'_1 z'_2 \cdots z'_{n'}, \quad z_i, z'_j \in Z, \tag{1.2.2}$$

can be rewritten

$$y_1 z^{k_1} y_2 z^{k_2} \cdots y_n z^{k_n} = y'_1 z^{k'_1} y'_2 z^{k'_2} \cdots y'_{n'} z^{k'_{n'}} \tag{1.2.3}$$

with $z_i = y_i z^{k_i}$, $z'_j = y'_j z^{k'_j}$, $y_i, y'_j \in Y - Z$, $k_i, k'_j \geq 0$. Since Y is a code Eq. (1.2.3) is trivial. This implies $y_1 = y'_1$, $k_1 = k'_1$, $y_2 = y'_2, \ldots$ and finally $n = n'$ and $z_i = z'_i$.

But we have $X \subset Z^*$ and $Z^* \subset Y^*$, which contradicts the minimality of the submonoid Y^*. Hence α is surjective, which implies that Y has fewer elements than X. $\blacksquare$

As an immediate consequence, of Theorem 1.2.5, there is the following corollary.

COROLLARY 1.2.6. *Each pair of words* $\{x, y\}$ *$(x, y \in A^*)$ is a code unless x and y are powers of a single word $z \in A^*$.*

Morphisms of free monoids play an essential role in the sequel. Let

$$\varphi : B^* \to A^*$$

be a morphism of free monoids. Clearly it is completely characterized by the images $\varphi(b) \in A^*$ of the letters $b \in B$. It is an isomorphism of B^* into A^* iff its restriction to B is injective and if the submonoid $\varphi(B^*)$ is a free submonoid of A^*.

A morphism $\varphi : B^* \to A^*$ is called *nonerasing* if $\varphi(B^+) \subset A^+$. If φ nonerasing, then for all $w \in B^*$,

$$|\varphi(w)| \geq |w|.$$

1.3. Conjugacy

A word $x \in A^*$ is said to be *primitive* if it is not a power of another word; that is, if $x \neq 1$ and $x \in z^*$ for $z \in A^*$ implies $x = z$.

PROPOSITION 1.3.1. *If*

$$x^n = y^m, \quad x, y \in A^*, n, m \geq 0,$$

there exists a word z such that $x, y \in z^*$.

In particular, for each word $w \in A^+$, there exists a unique primitive word x such that $w \in x^*$.

Proof. If $w = x^n = y^m$ with $x \neq y$ the set $\{x, y\}$ is not a code and, by the defect theorem (1.2.5) there exists a word $z \in A^*$ such that $x, y \in z^*$. If $w = x^n = y^m$ with x and y primitive, then there exists a word $z \in A^*$ such that $x = z^i$, $y = z^j$, $i, j \geqslant 0$. This implies $x = y = z$. ∎

PROPOSITION 1.3.2. *Two words* $x, y \in A^+$ *commute iff they are powers of the same word. More precisely the set of words commuting with a word* $x \in A^+$ *is a monoid generated by a single primitive word.*

Proof. Let z be the unique primitive word such that $x \in z^*$. Then if $xy = yx$ for $y \in A^+$, the set $\{x, y\}$ is not a code and there exists $t \in A^+$ such that $x, y \in t^*$. Then by Proposition 1.3.1, $t \in z^*$. Therefore the set of words commuting with x is generated by z. ∎

Two words x and y are said to be *conjugate* if there exist words $u, v \in A^*$ such that

$$x = uv, \qquad y = vu. \tag{1.3.1}$$

This is an equivalence relation on A^* since x is conjugate to y iff y can be obtained by a cyclic permutation of the letters of x. More precisely, let γ be the permutation of A^+ defined by

$$\gamma(ax) = xa, \qquad a \in A, \quad x \in A^*;$$

then the classes of conjugate elements are the orbits of γ.

PROPOSITION 1.3.3. *Let* $x, y \in A^*$ *and* z, t *be the primitive words such that* $x \in z^*$, $y \in t^*$. *Then* x *and* y *are conjugate iff* z *and* t *are also conjugate; in this case, there exists a unique pair* $(u, v) \in A^* \times A^+$ *such that* $z = uv, t = vu$.

Proof. Let $x = z^k$. If $x = rs$, there exists $u, v \in A^*$ such that $z = uv$, $r = z^{k_1}u$, $s = vz^{k_2}$ and $k_1 + k_2 + 1 = k$. Then the conjugate $y = sr$ of x can be written $y = t^k$ with $t = vu$. Moreover the pair (u, v) such that $z = uv, t = vu$ is unique since by Proposition 1.3.2 z has $|z|$ distinct conjugates. ∎

PROPOSITION 1.3.4. *Two words* $x, y \in A^+$ *are conjugate iff there exists a* $z \in A^*$ *such that*

$$xz = zy. \tag{1.3.2}$$

More precisely, equality (1.3.2) *holds iff there exist* $u, v \in A^*$ *such that*

$$x = uv, \qquad y = vu, \qquad z \in u(vu)^*. \tag{1.3.3}$$

Proof. If Eq. (1.3.3) holds, then (1.3.2) also holds. Conversely, if $xz = zy$, for $x, y \in A^+, z \in A^*$, we have for each $n \geq 1$

$$x^n z = zy^n. \tag{1.3.4}$$

Let n be such that $n|x| \geq |z| \geq (n-1)|x|$. Then we deduce from Eq. (1.3.4) that

$$z = x^{n-1}u, \qquad x = uv, \qquad vz = y^n. \tag{1.3.5}$$

Finally $y^n = vz = vx^{n-1}u$ is also equal to $(vu)^n$ and since $|y| = |x|$, we obtain $y = vu$, proving that Eq. (1.3.3) holds. ∎

It may be observed that, in accordance with the defect theorem, the equality $xz = zy$ implies $x, y, z \in \{u, v\}^*$, a submonoid with two generators.

The properties of conjugacy in A^* proved thus far can be viewed as particular cases of the properties of conjugacy in the free group on A (see Problem 1.3.1).

If $\mathrm{Card}(A) = k$ is finite, let us denote by $\psi_k(n)$ the number of classes of conjugates of primitive words of length n on the alphabet A. If w is a word of length n and if $w = z^q$ with z primitive and $n = qd$, then the number of conjugates of w is exactly d. Hence

$$k^n = \sum_{d \mid n} d\psi_k(n/d), \tag{1.3.6}$$

the sum running over the divisors of n. By Möbius inversion formula (see Problem 1.3.2) this is equivalent to:

$$\psi_k(n) = \frac{1}{n} \sum_{d \mid n} \mu(d)k^{n/d} \tag{1.3.7}$$

where μ is the *Möbius function* defined on $\mathbf{N} - 0$ as follows:

$$\mu(1) = 1,$$

$$\mu(n) = (-1)^i$$

if n is the product of i distinct primes and

$$\mu(n) = 0$$

if n is divisible by a square.

Proposition 1.3.1 admits the following refinement (Fine and Wilf 1965):

PROPOSITION 1.3.5. *Let x, $y \in A^*$, $n = |x|$, $m = |y|$, $d = gcd(n, m)$. If two powers x^p and y^q of x and y have a common left factor of length at least equal to $n + m - d$, then x and y are powers of the same word.*

Proof. Let u be the common left factor of length $n + m - d$ of x^p, y^q. We first suppose that $d = 1$ and show that x and y are powers of a single letter. We may assume that $n \leqslant m - 1$. It will be enough to show that the first $n - 1$ letters of u are equal. Denote by $u(i)$ the ith letter of u. By hypothesis, we have

$$u(i) = u(i + n), \qquad 1 \leqslant i \leqslant m - 1, \qquad (1.3.8)$$
$$u(j) = u(j + m), \qquad 1 \leqslant j \leqslant n - 1. \qquad (1.3.9)$$

Let $1 \leqslant i$, $j \leqslant m - 1$ and $j \equiv i + n \bmod m$. Then either $j = i + n$ or $j = i + n - m$. In the first case $u(i) = u(j)$ by (1.3.8). In the second case $u(j) = u(j + m)$ by (1.3.9) since $j = i + n - m \leqslant n - 1$. Therefore

$$u(i) = u(i + n) = u(j + m) = u(j).$$

Hence $u(i) = u(j)$ whenever $1 \leqslant i$, $j \leqslant m - 1$ and $j - i \equiv n \bmod m$. But since m, n are supposed to be relatively prime, any element of the set $\{1, 2, \ldots, m - 1\}$ is equal modulo m to a multiple of n. This shows that the first $m - 1$ letters of u are equal. In the general case, we consider the alphabet $B = A^d$ and, by the foregoing argument, x and y are powers of a single word of length d. ∎

Example 1.3.6. Consider the sequence of words on $A = \{a, b\}$ defined as follows: $f_1 = b$, $f_2 = a$ and

$$f_{n+1} = f_n f_{n-1}, \quad n \geqslant 2.$$

The sequence of the lengths $\lambda_n = |f_n|$ is the Fibonacci sequence. Two consecutive elements λ_n and λ_{n+1} for $n \geqslant 3$ are relatively prime. Let g_n be the left factor of f_n of length $\lambda_n - 2$ for $n \geqslant 3$. Then

$$g_{n+1} = f_{n-1}^2 g_{n-2}$$

for $n \geqslant 5$, as it may be verified by induction. We then have simultaneously

$$f_{n+1} \leqslant f_n^2, \quad g_{n+1} \leqslant f_{n-1}^3.$$

Therefore, for each $n \geqslant 5$, f_n^2 and f_{n-1}^3 have a common left factor of length $\lambda_n + \lambda_{n-1} - 2$. This shows that the bound given by Proposition 1.3.5

is optimal. For instance,

$$g_7 = \overbrace{a b a a b}^{f_5} \overbrace{a b a a b}^{f_5} a \quad \overbrace{}^{f_6}$$

1.4. Formal Series

Enumeration problems on words often lead to considering mappings of the free monoid into a ring. Such mappings may be viewed (and usefully handled) as finite or infinite linear combinations of words (see for instance Problem 1.4.2). This is the motivation for introducing the concept of a formal series.

Let K be a ring with unit; in the sequel K will be generally be the ring $\mathbb{Z}$ of all integers. A *formal series* (or series) with coefficients in K and variables in A is just a mapping of the free monoid A^* into K. The set of these series is denoted by $K\langle\langle A\rangle\rangle$.

For a series $\sigma \in K\langle\langle A\rangle\rangle$ and a word $w \in A^*$, the value of σ on w is denoted by $\langle\sigma, w\rangle$ and called the *coefficient* of w in σ; it is an element of K.

For a set $X \subset A^*$, we denote by $\mathbf{X}$ the *characteristic* series of X, defined by

$$\langle \mathbf{X}, x \rangle = 1 \quad \text{if } x \in X,$$
$$\langle \mathbf{X}, x \rangle = 0 \quad \text{if } x \notin X.$$

The operations of sum and product of two series $\sigma, \tau \in K\langle\langle A\rangle\rangle$ are defined by:

$$\langle \sigma + \tau, w \rangle = \langle \sigma, w \rangle + \langle \tau, w \rangle,$$
$$\langle \sigma\tau, w \rangle = \sum_{w = uv} \langle \sigma, u \rangle \langle \tau, v \rangle,$$

for any $w \in A^*$. These operations turn the set $K\langle\langle A\rangle\rangle$ into a ring. This ring has a unit that is the series $\mathbf{1}$, where 1 is the empty word.

A formal series $\sigma \in K\langle\langle A\rangle\rangle$ such that all but a finite number of its coefficients are zero is called a *polynomial*. The set $K\langle A\rangle$ of these polynomials is a subring of the ring $K\langle\langle A\rangle\rangle$. It is called the free (associative) K-algebra over A (see Problem 1.4.1). For each $\sigma \in K\langle\langle A\rangle\rangle$ and $\tau \in K\langle A\rangle$, we define

$$\langle \sigma, \tau \rangle = \sum_w \langle \sigma, w \rangle \langle \tau, w \rangle.$$

This is a bilinear map of $K\langle\langle A\rangle\rangle \times K\langle A\rangle$ in K.

The sum may be extended to an infinite number of elements with the following restriction: A family $(\sigma_i)_{i \in I}$ of series is said to be *locally finite* if for each $w \in A^*$, all but finitely many of the coefficients $\langle \sigma_i, w \rangle$ are zero.

If $(\sigma_i)_{i \in I}$ is a locally finite family of series, the sum

$$\sigma = \sum_{i \in I} \sigma_i$$

is well defined since for each $w \in A^*$, the coefficient $\langle \sigma, w \rangle$ is the sum of a finite set of nonzero coefficients $\langle \sigma_i, w \rangle$.

In particular, the family $(\mathbf{w})_{w \in A^*}$ is locally finite, and this allows to write for any $\sigma \in K\langle\langle A \rangle\rangle$

$$\sigma = \sum_{w \in A^*} \langle \sigma, w \rangle \mathbf{w},$$

or, by identifying w with $\mathbf{w}$,

$$\sigma = \sum_{w \in A^*} \langle \sigma, w \rangle w.$$

This is the usual notation for formal series in one variable:

$$\sigma = \sum_{n \geq 0} \sigma_n a^n$$

with $\sigma_n = \langle \sigma, a^n \rangle$.

Let σ be a series such that $\langle \sigma, 1 \rangle = 0$; the family $(\sigma^i)_{i \geq 0}$ is then locally finite since $\langle \sigma^i, w \rangle = 0$ for $i \geq |w| + 1$. This allows us to define the new series

$$\sigma^* = 1 + \sigma + \sigma^2 + \cdots,$$

which is called the *star* of σ. It is easy to verify the following:

PROPOSITION 1.4.1. *Let $\sigma \in K\langle\langle A \rangle\rangle$ be such that $\langle c, 1 \rangle = 0$. The series σ^* is the unique series such that*:

$$\sigma^*(1 - \sigma) = (1 - \sigma)\sigma^* = 1.$$

Following is a list of statements relating the operations in $K\langle\langle A \rangle\rangle$ with the operations on the subsets of A^* when K is assumed to be of characteristic zero.

PROPOSITION 1.4.2. *For two subsets X, Y of A^*, one has*

(i) *let $Z = X \cup Y$. Then $\mathbf{Z} = \mathbf{X} + \mathbf{Y}$ iff $X \cap Y = \emptyset$,*
(ii) *let $Z = XY$. Then $\mathbf{Z} = \mathbf{XY}$ iff $xy = x'y' \Rightarrow x = x'$, $y = y'$, for $x, x' \in X$, $y, y' \in Y$,*
(iii) *let $X \subset A^+$, and $P = X^*$. Then $\mathbf{P} = \mathbf{X}^*$ iff X is a code.*

The proof is left to the reader as an exercise.

Notes

The terminology used for words presents some variations in the literature. Some authors call *subword* what is called here *factor* the term *subword* is reserved for another use (see Chapter 6). Some call *prefix* or *initial segment* what we call a *left factor*. Also, the *empty word* is often denoted by ε instead of 1.

General references concerning free submonoids are Eilenberg 1974 and Lallement 1977; Proposition 1.2.3 was known to Schützenberger (1956) and to Cohn (1962). The defect theorem (Theorem 1.2.5) is virtually folklore; it has been proved under various forms by several authors (see Lentin 1972; Makanin 1976; Ehrenfeucht and Rozenberg 1978). The proof given here is from Berstel et al. 1979, where some generalizations are discussed.

The results of Section 1.3 are also mainly common knowledge. For further references see Chapters 8 and 9.

The standard reference for Section 1.4 is Eilenberg 1974.

Problems

Section 1.1

1.1.1. (Levi's lemma). A monoid M is free iff there exists a morphism λ of M into the monoid $\mathbb{N}$ of additive integers such that $\lambda^{-1}(0) = 1_M$ and if for any $x, y, z, t \in M$

$$xy = zt$$

implies the existence of a $u \in M$ such that either $x = zu, uy = t$ or $xu = z, y = ut$.

1.1.2. Let A be an alphabet and $\bar{A} = \{\bar{a} \mid a \in A\}$ be a copy of A. Consider in the free monoid over the set $A \cup \bar{A}$ the congruence generated by the relations

$$a\bar{a} = \bar{a}a = 1, \qquad a \in A.$$

a. Show that each word has a unique representative of minimal length, called a *reduced word*.

b. Show that the quotient of $(A \cup \bar{A})^*$ by this congruence is a group F; the inverse of the reduced word w is denoted by $\bar{w}$.

c. Show that for any mapping α of A into a group G, there exists a unique morphism φ of F onto G making the following diagram commutative:

F is called the *free group* over A (see Magnus, Karass, and Solitar 1976 or Hall 1959 or Lyndon and Schupp 1977). Henceforth in problems about free groups,

$$\rho: F \to (A \cup \overline{A})^*,$$

denotes the mapping associating to each element of F the unique reduced word representing it.

*1.1.3. Let $\varphi: A^* \to A^*$ be the mapping assigning to each word $w \in A^*$ the longest word that is both a proper left and a proper right factor of w.

  a. Let $w = a_1 a_2 \cdots a_n$ and denote $\varphi(i) = j$ instead of $\varphi(a_1 \cdots a_i) = a_1 \cdots a_j$; show that the following algorithm allows computation of φ:

    1. $\varphi(1) \leftarrow 0$;
    2. *for* $i \leftarrow 2$ *until* n *do*
          begin
    3. $j \leftarrow \varphi(i-1)$;
    4.   *while* $j > 0$ *and* $a_i \neq a_{j+1}$ *do* $j \leftarrow \varphi(j)$;
    5.   *if* $j = 0$ *and* $a_i \neq a_{j=1}$ *then* $\varphi(i) \leftarrow 0$
    6.   *else* $\varphi(i) \leftarrow j + 1$;
          end

  (For the notations concerning algorithms, see Aho, Hopcroft, and Ullman 1974.)

  b. Show that the number of successive comparisons of two letters of the word w in performing the foregoing algorithm does not exceed $2n$. (*Hint:* Note that the variable j can be increased at most n times by one unit.)

  c. Show that the foregoing algorithm can be used to test whether a word $u \in A^+$ is a factor of a word $v \in A^+$. (*Hint:* Apply the algorithm of (a) to the word $w = uv$.)

This is called a *string-matching algorithm* (see Knuth, Morris, and Pratt 1977).

Section 1.2

*1.2.1. Let F be the free group over the set A and H be a subgroup of F.

  a. Show that it is possible to choose a set Q of representatives of the right cosets of H in F such that the set $\rho(Q)$ of reduced words representing Q contains all its left factors. Such a set Q is called a *Schreier system* for H.

  b. Let Q be a Schreier system for H and

$$X = \{ pa\overline{q} \mid p, q \in Q, a \in A, pa \in (H-1)q \};$$

Show that X generates H.

    c. Show that each $pa\bar{q} \in X$ is reduced as written and that, in the product of two elements of $X \cup \bar{X}$ the letters a in the triple (p, a, q) never cancel unless the whole product does.

    d. Deduce from (a), (b), and (c) that any subgroup H of a free group F is free and that if H is of finite index d in F, then it is isomorphic with a free group on r generators with

$$r - 1 = d(k - 1), \quad k = \mathrm{Card}(A)$$

(Schreier's formula; see the references of Problem 1.1.2).

1.2.2. A submonoid N of A^* is generated by a prefix iff it satisfies:

$$m, mn \in N \Rightarrow n \in N$$

for all $m, n \in A^*$. Such a submonoid is called (right) *unitary*.

1.2.3. Let P be the set of words

$$P = \{ w\tilde{w} \mid w \in A^* \}.$$

Then P is the set of *palindromes* (i.e., $u = \tilde{u}$) of even length. Show that the submonoid P^* is right and left unitary.
(*Hint:* Let Π be the basis of P^*; show that Π is prefix.) (See Knuth, Morris and Pratt 1977.)

1.2.4. Let $\theta : A^* \to B^*$ be a morphism and $P \subset B^*$ be a free submonoid of B^*. Show that $\theta^{-1}(P)$ is a free submonoid of A^*.

Section 1.3

1.3.1. Show that two words $x, y \in A^*$ are conjugate iff they are conjugate in the free group F over A—that is, iff there exists an element g of F such that

$$x = g y g^{-1},$$

(Identify A^* to a subset of F.)

1.3.2. (*Möbius inversion formula*) Let μ be the Möbius function; show that

$$\sum_{d \mid n} \mu(d) = \begin{vmatrix} 1 & \text{if } n = 1, \\ 0 & \text{if } n \geqslant 2. \end{vmatrix}$$

Deduce from this that two functions φ, ψ of $\mathbb{N} - 0$ in $\mathbb{Z}$ are related by

$$\sum_{d \mid n} \psi(d) = \varphi(n)$$

iff

$$\sum_{d \mid n} \mu(d)\varphi(n/d) = \psi(n).$$

1.3.3. Show directly (without using the defect theorem, that is) that if $\{x, y\}$ is not a code, then x and y are powers of a single word.

1.3.4. (Problem 1.1.3) Show that $\varphi(w) = u$ iff

$$w = (st)^{k+1}s, u = (st)^k s, \quad k \geqslant 0, s, t \in A^*$$

with $|s|$ minimal. Deduce that the algorithm of Problem 1.1.3 allows computation of the primitive word such that $w = v^n, n \geqslant 1$ (*Hint:* Use Proposition 1.3.5.)

1.3.5. Let $w = a_1 a_2 \cdots a_n, n \geqslant 1, a_i \in A$. For $1 \leqslant i \leqslant n$, let $\psi(i)$ be the greatest integer $j \leqslant i - 1$ such that

$$a_1 a_2 \cdots a_{j-1} = a_{i-j+1} \cdots a_{i-1}, \quad a_i \neq a_j.$$

with $\psi(i) = 0$ if no such integer j exists.

a. Show that the following algorithm computes ψ:

   1. $\psi(1) \leftarrow 0$;
   2. $i \leftarrow 1; j \leftarrow 0$;
   3. *while* $i < n$ *do*
        begin
   4.   *while* $j > 0$ *and* $a_i \neq a_j$ *do* $j \leftarrow \psi(j)$;
   5. $i \leftarrow i + 1$; $j \leftarrow j + 1$;
   6.   *if* $a_i = a_j$ *then* $\psi(i) \leftarrow \psi(j)$ *else* $\psi(i) \leftarrow j$;
        end

(*Hint:* Show that the value of the variable j at line 6 is $\varphi(i-1)+1$ wheres φ is as in Problem 1.1.3.)

b. Show that the algorithm of problem part (a) can be used to test whether a word u is a factor of a word v.

c. Show that the number of consecutive times the *while* loop of line 4 may be executed does not exceed the integer r such that

$$\lambda_{r+3} \leqslant n < \lambda_{r+4}$$

where λ_r is the rth term of the Fibonacci sequence. Show, using the sequence of Example 1.3.6 that this bound can be reached. (See Knuth, Morris and Pratt 1978; Duval 1981.)

Section 1.4

1.4.1. For any mapping α of A into an associative K-algebra R, there exists a unique morphism φ of $K\langle A \rangle$ into R such that the following diagram is commutative:

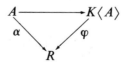

1.4.2. Let $W \subset A^+$ and

$$P = A^* - A^*WA^*$$

be the set of words having no factor in W. Let for each $u \in W$,

$$X_u = A^*u - A^*WA^+$$

be the set of words having u as a right factor but no other factor in W. For each $u, v \in W$, let $R_{u,v}$ be the finite set

$$R_{u,v} = \{t \in A^+ - A^*v \mid ut \in A^*v\}.$$

a. Show that the following equalities hold in $\mathbb{Z}\langle\langle A \rangle\rangle$:

$$1 + \mathbf{P}A = \mathbf{P} + \sum_{u \in W} \mathbf{X}_u, \tag{a.1}$$

and for each $u \in W$,

$$\mathbf{P}u = \mathbf{X}u + \sum_{v \in W} \mathbf{X}\mathbf{R}_{v,u} \tag{a.2}$$

b. Show that the system of equalities (a.1) and (a.2), for $u \in W$, allows computation of P.

c. Show that the formal series

$$\lambda = \sum_{n \geqslant 0} \lambda_n z^n$$

with $\lambda_n = \text{Card}(A^n \cap P)$, is rational. (*Hint:* Use the morphism of $\mathbb{Z}\langle\langle A \rangle\rangle$ onto $\mathbb{Z}\langle\langle z \rangle\rangle$ sending $a \in A$ on z.)

d. Apply the foregoing method to show that for $W = \{aba\}$ one has

$$\lambda_n = 2\lambda_{n-1} - \lambda_{n-2} + \lambda_{n-3}, \quad n \geqslant 3.$$

(See Schützenberger 1964; for a general reference concerning linear equations in the ring $\mathbb{Z}\langle\langle A \rangle\rangle$, see Eilenberg 1974.)

CHAPTER 2

Square-Free Words and Idempotent Semigroups

2.0. Introduction

The investigation of words includes a series of combinatorial studies with rather surprising conclusions that can be summarized roughly by the following statement: Each sufficiently long word over a finite alphabet behaves locally in a regular fashion. That is to say, an arbitrary word, subject only to the constraint that it be sufficiently long, possesses some regularity. This claim becomes meaningful only if one specifies the kind of regularities that are intended, of course. The discovery and the analysis of these *unavoidable regularities* constitute a major topic in the combinatorics of words. A typical example is furnished by van der Waerden's theorem.

It should not be concluded that any sufficiently long word is globally regular. On the contrary, the existence of unavoidable regularities leads to the dual question of avoidable regularities: properties not automatically shared by all sufficiently long words. For such a property there exist infinitely many words (finiteness of the alphabet is supposed) that do not satisfy it. The present chapter is devoted mainly to the study of one such property.

A *square* is a word of the form uu, with u a nonempty word. A word contains a square if one of its factors is a square; otherwise, the word is called *square-free*. For instance, *abcacbacbc* contains the square *acbacb*, and *abcacbabcb* is square-free. The answer to the question of whether every sufficiently long word contains a square is no, provided the alphabet has at least three letters. As will be shown, the existence of infinitely many square-free words is equivalent to the existence of a square-free word that is infinite (on the right). The formalism of infinite words has the advantage of allowing concise descriptions. Furthermore, infinite iteration of a morphism is a natural and simple way to construct infinite words, and this method applies especially to the construction of infinite square-free words.

We start with the investigation of a famous infinite word, called after its discoverers the word of Thue–Morse. This word contains squares, but it is cube-free and even has stronger properties. Then we turn to the study of infinite square-free words. A simple coding of the Thue–Morse infinite word gives an example of an infinite square-free word. We then establish a general result of Thue that gives other infinite square-free words.

A more algebraic framework can be used for the theory of square-free words. Consider the monoid $M = A^* \cup 0$ obtained by adjoining a zero to the free monoid A^*. Next consider the congruence over M generated by the relations

$$uu \approx 0 \qquad (u \in A^+).$$

The fact that there exist infinitely many square-free words can be rephrased: The quotient monoid $M/\approx$ is infinite, provided A has at least three letters. A natural analogue is to consider the free idempotent monoid, that is, the quotient of A^* by the congruence generated by

$$uu \sim u \qquad (u \in A^+).$$

We will show, in contrast to the previous result, that for each finite alphabet A, the quotient monoid $A^*/\sim$ is finite.

Many results, extensions, and generalizations concerning the problems just sketched are not included in the text. They are stated as exercises or briefly mentioned in the Notes, which also contain some bibliographic remarks.

2.1. Preliminaries

Before defining infinite words, let us fix some notations concerning distinct occurrences of a word as a factor in a given word. Let A be an alphabet, $w \in A^+$. Let u be a nonempty word having two distinct occurrences as a factor in w. Then there are words $x, y, x', y' \in A^*$ such that

$$w = xuy = x'uy', \quad x \neq x'.$$

These two occurrences of u either overlap or are consecutive or are disjoint. More precisely, we may suppose $|x| < |x'|$. Then three possibilities arise (see Figure 2.1).

(i) $|x'| > |xu|$. In this case, $x' = xuz$ for some $z \in A^+$, and $w = xuzuy'$. The occurrences of u are *disjoint*.

(ii) $|x'| = |xu|$. This implies that $x' = xu$, and consequently $w = xuuy'$ contains a square. The occurrences of u are *adjacent*.

(iii) $|x'| < |xu|$. The two occurrences of u are said to *overlap*. The following lemma gives a more precise description of this case.

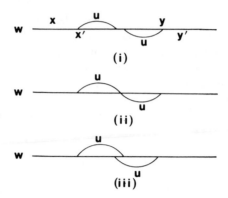

Figure 2.1. Two occurrences of u in w: (i) disjoint occurrences, (ii) adjacent occurrences, (iii) overlapping occurrences.

LEMMA 2.1.1. *Let w be a word; then w contains two overlapping occurrences of a word $u \neq 1$ iff w contains a factor of the form $avava$, with a a letter and v a word.*

Proof. Assume first $w = xuy = x'uy'$, where the occurrences of u overlap. Then $|x| < |x'| < |xu| < |x'u|$. Consequently

$$x' = xs, \qquad xu = x'z, \qquad x'u = xut$$

for some nonempty words s, z, t, whence

$$u = sz = zt. \tag{2.1.1}$$

Let a be the first letter of s, and therefore also of z by Eq. (2.1.1). Set $s = av, z = az'$. Then by (2.1.1) $u = avaz'$ and

$$w = xsuy' = xavavaz'y'.$$

Conversely, if $avava$ is a factor of w, then $u = ava$ clearly has two overlapping occurrences in w. ∎

A word of the form $avava$, with a a letter, is said to *overlap*. Thus, according to the lemma, a word has two overlapping occurrences of a word iff it contains an overlapping factor.

We now turn to the definition of infinite words. Let A be an alphabet. An *infinite word* on A is a function

$$\mathbf{a}\colon \mathbb{N} \to A.$$

We use the following notation

$$\mathbf{a} = \mathbf{a}(0)\mathbf{a}(1)\cdots\mathbf{a}(n)\cdots,$$

and also

$$\mathbf{a} = a_0 a_1 \cdots a_n \cdots,$$

where $a_n = \mathbf{a}(n)$ is a letter. The left factor of length $k \geqslant 0$ of $\mathbf{a}$ is

$$\mathbf{a}^{[k]} = a_0 a_1 \cdots a_{k-1}.$$

For $u \in A^*$, we write $u < \mathbf{a}$ whenever $u = \mathbf{a}^{[k]}$ for $k = |u|$. Then clearly $\mathbf{a} = u\mathbf{b}$ where $\mathbf{b}(m) = \mathbf{a}(m + k)$ for all $m \geqslant 0$. A *factor* of u is any word in A^* that occurs in $\mathbf{a}$. In the sequel, by a word we always mean a finite word.

Infinite words are useful when one deals with properties P of (finite) words having a special feature, namely that $P(xuy)$ implies $P(u)$ for all words x, u, y. In other terms, if L_P is the set of words for which P holds, then L_P contains the factors of its elements. Note that this holds for the set of square-free words. When P satisfies this condition we say that P is *stable for factors*. Given an infinite word $\mathbf{a}$, we say that $\mathbf{a}$ has the property P if each factor of $\mathbf{a}$ satisfies P. Thus it is meaningful to speak about infinite square-free words.

LEMMA 2.1.2. *Let A be a finite alphabet and let P be a property of elements of A that is stable for factors. Then the two following conditions are equivalent:*

(i) *The set L_P of words w in A such that $P(w)$ is infinite.*
(ii) *There exists an infinite word on A with property P.*

A particular case is the assertion mentioned in the introduction, namely that the existence of infinitely many square-free words is equivalent to the existence of an infinite square-free word.

Proof. Clearly (ii) implies (i). Conversely, if $L = L_P$ is infinite, the finiteness of A implies that infinitely many words in L start with a same letter, say a_0. Set $L_0 = L \cap a_0 A^*$. Assume by induction that there are letters $a_0, a_1, \ldots, a_n$ such that $L_n = L \cap a_0 a_1 \cdots a_n A^*$ is infinite. Then among the sets $(L \cap a_0 a_1 \cdots a_n b A^*)_{b \in A}$ at least one is infinite. Choose one letter a_{n+1} such that $L \cap a_0 a_1 \cdots a_n a_{n+1} A^*$ is infinite.

Thus there exists a sequence $a_0, a_1, \ldots, a_n, \ldots$ of letters in A such that $L \cap a_0 a_1 \cdots a_n A^*$ is infinite for each $n \geqslant 0$. Define $\mathbf{a} \colon \mathbb{N} \to A$ by $\mathbf{a}(n) = a_n$. Then each factor of $\mathbf{a}$ is a factor of a word in L, thus is itself in L. ∎

Sometimes a simpler method can be applied to construct infinite words from finite ones. (Note that the proof of the previous lemma gives such a construction.)

Let $w_0, w_1, \ldots, w_n, \ldots$ be a sequence of words in A^* of unbounded length such that each w_{n-1} is a left factor of w_n. Then define an infinite word $\mathbf{a}$ on A by

$$\mathbf{a}^{[k]} = w_n, \qquad k = |w_n|, \qquad n \geq 0.$$

The definition is consistent because $\mathbf{a}^{[k]}$ is a left factor of all $w_m, m \geq n$. The infinite word defined in this way is called the *limit* of $(w_n)_{n \geq 0}$ and is denoted by

$$\mathbf{a} = \lim w_n.$$

Consider the following important special case. Let

$$\alpha: A^* \to A^*$$

be a morphism verifying

   (i) $\alpha(a) \neq 1$ for $a \in A$,   (2.1.2)
   (ii) there exists a letter a_0 such that

$$\alpha(a_0) = a_0 u \quad \text{for some} \quad u \in A^+.$$ (2.1.3)

Then for each $n \geq 0$,

$$\alpha^{n+1}(a_0) = \alpha^n(a_0 u) = \alpha^n(a_0)\alpha^n(u).$$

Thus each $\alpha^n(a_0)$ is a proper left factor of $\alpha^{n+1}(a_0)$, and therefore the limit of the sequence $(\alpha^n(a_0))_{n \geq 0}$ exists. We denote this limit by $\alpha^\omega(a_0)$:

$$\alpha^\omega(a_0) = \lim \alpha^n(a_0),$$

and we say that it is obtained by *iterating* α on a_0.

With these notations α can be extended to infinite words by setting, for $\mathbf{b} = b_0 b_1 \cdots b_n \cdots$

$$\alpha(\mathbf{b}) = \alpha(b_0)\alpha(b_1) \cdots \alpha(b_n) \cdots.$$

Condition (i) ensures that $\alpha(\mathbf{b})$ is indeed an infinite word. Observe that

$$\alpha(\mathbf{a}) = \mathbf{a} \quad \text{for} \quad \mathbf{a} = \alpha^\omega(a_0).$$ (2.1.4)

In other terms, $\mathbf{a}$ is a fixed point for α. Indeed set $\mathbf{b} = \alpha(\mathbf{a})$. For each left factor u of $\mathbf{a}$, the word $\alpha(u)$ is a left factor of $\alpha(\mathbf{a})$. Thus each $\alpha^n(a_0), n \geq 1$, is a left factor of $\mathbf{b}$, and $\mathbf{b}$ starts with a_0 by (ii). Consequently $\mathbf{b} = \lim \alpha^n(a_0) = \mathbf{a}$; this proves Eq. (2.1.4).

2.2. The Infinite Words of Thue–Morse

In this section a special infinite word is defined and its properties are studied. The main result is that this infinite word has no overlapping factor.

In this section A denotes the fixed two-letter alphabet $A = \{a, b\}$. Define a morphism

$$\mu \colon A^* \to A^*$$

by

$$\mu(a) = ab, \qquad \mu(b) = ba.$$

Then μ satisfies conditions (2.1.2), (2.1.3) for $a_0 = a$ and also for $a_0 = b$. Consequently, iteration of μ on a and on b yields two infinite words

$$\mathbf{t} = \mu^\omega(a), \qquad \bar{\mathbf{t}} = \mu^\omega(b).$$

By definition, $\mathbf{t}$ is the *infinite word of Thue – Morse*. Computation gives

$$\mu(a) = ab \qquad\qquad \mu(b) = ba$$
$$\mu^2(a) = abba \qquad\qquad \mu^2(b) = baab$$
$$\mu^3(a) = abbabaab \qquad\qquad \mu^3(b) = baababba$$

$$\mathbf{t} = abbabaabbaababbabaabbaabbabaab\cdots$$
$$\bar{\mathbf{t}} = baababbaabbabaababbabaabbaababba\cdots$$

There are several properties relating the words $\mu^n(a)$, $\mu^n(b)$, $n \geq 0$. Consider the morphism

$$w \mapsto \bar{w}$$

defined by

$$\bar{a} = b, \qquad \bar{b} = a$$

Thus $\bar{w}$ is obtained from w by replacing each a by b and conversely. Of course $\bar{\bar{w}} = w$.

PROPOSITION 2.2.1. *Define $u_0 = a$, $v_0 = b$ and for $n \geq 0$*

$$u_{n+1} = u_n v_n, \qquad v_{n+1} = v_n u_n.$$

Then for all $n \geq 0$

(i) $u_n = \mu^n(a)$, $\quad v_n = \mu^n(b)$.
(ii) $v_n = \bar{u}_n$, $\quad u_n = \bar{v}_n$.
(iii) u_{2n}, v_{2n} *are palindromes and* $\tilde{u}_{2n+1} = v_{2n+1}$.

Proof. The proofs are by induction. The initial step is always clear. Formula (i) follows from

$$u_{n+1} = u_n v_n = \mu^n(a)\mu^n(b) = \mu^{n+1}(a),$$
$$v_{n+1} = v_n u_n = \mu^n(b)\mu^n(a) = \mu^{n+1}(b);$$

next (ii) follows from

$$v_{n+1} = v_n u_n = \bar{u}_n \bar{v}_n = \overline{u_n v_n} = \bar{u}_{n+1}, \bar{v}_{n+1} = \bar{\bar{u}}_{n+1} = u_{n+1};$$

finally for (iii), observe that for $k > 0$

$$\tilde{u}_k = (u_{k-1} v_{k-1})^{\tilde{}} = \tilde{v}_{k-1} \tilde{u}_{k-1}.$$

If k is odd (resp. even) this implies

$$\tilde{u}_k = v_{k-1} u_{k-1} = v_k \; (\text{resp. } \tilde{u}_k = u_{k-1} v_{k-1} = u_k). \qquad \blacksquare$$

There exists an interesting definition of

$$\mathbf{t} = t_0 t_1 \cdots t_n \cdots$$

that is independent of the morphism μ. First let, for $n \geqslant 0, d_2(n)$ be the number of 1's in the binary expansion of n. Then we have the following proposition.

PROPOSITION 2.2.2. *For each $n \geqslant 0$,*

$$t_n = \begin{cases} a & \text{if } d_2(n) \equiv 0 \quad mod\,2 \\ b & \text{if } d_2(n) \equiv 1 \quad mod\,2 \end{cases} \qquad (2.2.1)$$

Proof. Note that by (2.1.4) we have

$$\mathbf{t} = \mu(\mathbf{t}) = \mu(t_0)\mu(t_1) \cdots \mu(t_n) \cdots$$

and therefore $\mu(t_n) = t_{2n} t_{2n+1}$ for $n \geqslant 0$. By the definition of μ, this implies

$$t_{2n} = t_n, \qquad t_{2n+1} = \bar{t}_n \quad (n \geqslant 0). \qquad (2.2.2)$$

Formula (2.2.1) holds for $n = 0$. Thus let $n > 0$. If $n = 2m$, then $t_n = t_m$ by (2.2.2), and $d_2(n) = d_2(m)$. Thus (2.2.1) holds in this case. If $n = 2m + 1$, then $t_n = \bar{t}_m$ and $d_2(n) \equiv 1 + d_2(m) \, mod\,2$. Therefore (2.2.1) holds in this case too. $\qquad \blacksquare$

The inspection of **t** shows that **t** is not square-free. However, we will prove the following:

THEOREM 2.2.3. *The infinite word t has no overlapping factor.*

COROLLARY 2.2.4. *The infinite word t is cube-free.*

The proof of the theorem uses two lemmas.

LEMMA 2.2.5. *Let $X = \{ab, ba\}$; if $x \in X^*$, then $axa \notin X^*$ and $bxb \notin X^*$.*

Proof. By induction on $|x|$. If $|x| = 0$, then indeed $aa, bb \notin X^*$. Let $x \in X^*$, $x \neq 1$ and suppose $u = axa \in X^*$ (the case $bxb \in X^*$ is similar). Then $u = x_1 x_2 \cdots x_r$, with $x_1, \ldots, x_r \in X$; consequently $x_1 = ab$ and $x_r = ba$. Thus $u = abyba$ with $y = x_2 \cdots x_{r-1} \in X^*$. But now by induction $x = byb$ is not in X^*, contrary to the assumption. ∎

LEMMA 2.2.6. *Let $w \in A^+$. If w has no overlapping factor, then $\mu(w)$ has no overlapping factor.*

Proof. Assume that $\mu(w)$ has an overlapping factor for some $w \in A^*$. We show that w also has an overlapping factor.

By asumption, there are $x, v, y \in A^*$, $c \in A$ with

$$\mu(w) = xcvcvcy$$

Note that $|cvcvc|$ is odd, but $\mu(w) \in X^*$ with $X = \{ab, ba\}$: therefore $|\mu(w)|$ is even and $|xy|$ is odd. Thus

- Either: $|x|$ is even, and $x, cvcv, cy \in X^*$,
- Or: $|x|$ is odd, and $xc, vcvc, y \in X^*$.

This implies that $|v|$ is odd, since otherwise we get from $cvcv \in X^*$ (resp. $vcvc \in X^*$) that both v, cvc are in X^*, which contradicts Lemma 2.2.5.

In the case $|x|$ is even, it follows that cv is in X^* and $w = rsst$ with $\mu(r) = x, \mu(s) = cv, \mu(t) = cy$. But then s and t start with the same letter c and ssc is an overlapping factor in w.

In the case $|x|$ is odd, similarly $vc \in X^*$, and $w = rsst$ with $\mu(r) = xc, \mu(s) = vc, \mu(t) = y$. Here r and s end with c and css is an overlapping factor in w. ∎

Proof of Theorem 2.2.3. Assume that **t** has an occurrence of an overlapping factor. Then it occurs in a left factor $\mu^k(a)$ for some $k > 0$. On the other hand, since a has no overlapping factor, by iterated application of Lemma 2.2.6 no $\mu^n(a)$ $(n \geq 0)$ has an overlapping factor. Contradiction. ∎

2.3. Infinite Square-Free Words

The infinite word of Thue–Morse has square factors. In fact, the only square-free words over two letters a and b are

$$a, b, ab, ba, aba, bab.$$

On the contrary, there exist infinite square-free words over three letters. This will now be demonstrated.

As before let $A = \{a, b\}$, and let $B = \{a, b, c\}$. Define a morphism

$$\delta \colon B^* \to A^*$$

by setting

$$\delta(a) = a, \qquad \delta(b) = ab, \qquad \delta(c) = abb$$

For any infinite word $\mathbf{b}$ on B,

$$\delta(\mathbf{b}) = \delta(b_0)\delta(b_1)\cdots\delta(b_n)\cdots$$

is a well-defined infinite word on A starting with the letter a. Conversely, consider an infinite word $\mathbf{a}$ on A without overlapping factors and starting with a. Then $\mathbf{a}$ can be factored as

$$\mathbf{a} = y_0 y_1 \cdots y_n \cdots \tag{2.3.1}$$

with each $y_n \in \{a, ab, abb\} = \delta(B)$. Indeed, each a in $\mathbf{a}$ is followed by at most two b since bbb is overlapping, and then followed by a new a. Moreover, the factorization (2.3.1) is unique. Thus there exists a unique infinite word $\mathbf{b}$ on B such that $\delta(\mathbf{b}) = \mathbf{a}$.

THEOREM 2.3.1. *Let $\mathbf{a}$ be an infinite word on A starting with a, and without overlapping factor, and let $\mathbf{b}$ be the infinite word over B such that $\delta(\mathbf{b}) = \mathbf{a}$; then $\mathbf{b}$ is square-free.*

Proof. Assume the contrary. Then $\mathbf{b}$ contains a square, say uu. Let d be the letter following uu in one of its occurrences in $\mathbf{b}$. Then $\delta(uud)$ is a factor of $\mathbf{a}$. Since $\delta(u) = av$ for some $v \in A^*$ and $\delta(d)$ starts with a, $\mathbf{a}$ contains the factor $avava$. Contradiction.   ∎

By applying the theorem to the Thue–Morse word $\mathbf{t}$, we obtain an infinite square-free word $\mathbf{m}$ over the three letter alphabet B such that $\delta(\mathbf{m}) = \mathbf{t}$. This infinite word is

$$\mathbf{m} = abcacbabcbacabcacbacabcbabcacbabcbacabcbabc\cdots$$

Note that the converse of Theorem 2.3.1 is false: There are square-free infinite words **b** over B such that $\delta(\mathbf{b})$ has overlapping factors (see Problem 2.3.7). There are several alternative ways to obtain the word **m**. We quote just one.

PROPOSITION 2.3.2. *Define a morphism* $\varphi: B^* \to B^*$ *(with* $B = \{a, b, c\}$*) by* $\varphi(a) = abc, \varphi(b) = ac, \varphi(c) = b.$ *Then* $\mathbf{m} = \varphi^{\omega}(a).$

The proof is left as an exercise.

There exist other constructions that allow one to obtain more systematically infinite square-free words. We now present one of them. In the sequel of this paragraph, $A, B \cdots$ are again arbitrary alphabets.

First we introduce a new notion. A morphism $\alpha: A^* \to B^*$ is *square-free* if $\alpha(A) \neq \{1\}$ and if $\alpha(w)$ is a square-free word for each square-free word w. Thus a square-free morphism preserves square-free words. The first condition is present simply to avoid uninteresting discussions on the square-freeness of the empty word. A square-free morphism α from A^* into itself produces by iteration only square-free words, when one starts with a square-free word, or simply with a letter. Thus a square-free morphism usually gives an infinite set of square-free words. Note that the morphism φ of Proposition 2.3.2 is *not* square-free since

$$\varphi(abc) = ab\underline{caca}bc$$

contains a square. The following theorem gives sufficient conditions for a morphism to be square-free.

THEOREM 2.3.3. *Let* $\alpha: A^* \to B^*$ *be a morphism with* $\alpha(A) \neq \{1\}$ *such that*

(i) $\alpha(u)$ *is square-free for each square-free word of length* ≤ 3,
(ii) *No* $\alpha(a)$ *is a proper factor of an* $\alpha(b)$ *(* a, b *in* A *).*

Then α *is a square-free morphism.*

Proof. First we note that $\alpha(a) \neq 1$ for each $a \in A$; otherwise if $\alpha(a) = 1$ let $b \in A$ be a letter with $x = \alpha(b) \neq 1$. Then bab is square-free, but $\alpha(bab) = xx$ violates condition (i). Next α is injective on A: if $\alpha(a) = \alpha(b)$, then $\alpha(ab)$ is a square, consequently $a = b$ by (i). Furthermore, $X = \alpha(A)$ is a biprefix code by (ii). Now we prove the following claim.

Claim: If $\alpha(a_1 a_2 \cdots a_n) = x\alpha(a)y$ for $a, a_i \in A, x, y \in B^*$, then $a = a_j$ for some j, $x = \alpha(a_1 \cdots a_{j-1}), y = \alpha(a_{j+1} \cdots a_n)$.

The claim is clear for $n=1$ by (ii). Arguing by induction on n, assume $n>1$. If

$$|x\alpha(a)| \leqslant |\alpha(a_1 a_2 \cdots a_{n-1})|$$

or

$$|\alpha(a)y| \leqslant |\alpha(a_2 \cdots a_n)|,$$

the claim follows by the induction hypothesis. Thus, we may assume that both

$$|x\alpha(a)| > |\alpha(a_1 a_2 \cdots a_{n-1})|$$

and

$$|\alpha(a)y| > |\alpha(a_2 \cdots a_n)|.$$

Consequently, y is a proper right factor of $\alpha(a_n)$, and x is a proper left factor of $\alpha(a_1)$:

$$\alpha(a_1) = xu, \qquad \alpha(a_n) = vy$$

for some u, v in B^+, and

$$\alpha(a) = u\alpha(a_2) \cdots \alpha(a_{n-1})v.$$

By (ii), this implies $n=2$ and $\alpha(a) = uv$.

The words $\alpha(a_1 a) = xuuv$ and $\alpha(aa_n) = uvvy$ are not square-free. According to (i), $a_1 = a = a_n$, whence

$$xu = uv = vy.$$

The first equation shows that $|x| = |v|$. In view of $xu = vy$, it follows that $x = v$. Consequently $vu = uv$. By a result of Chapter 1, $\alpha(a) = uv$ is not a primitive word and thus is not square-free. This contradicts condition (i) and proves the claim.

Now we prove the theorem. Assume the conclusion is false. Then there is a shortest square-free word $w \in A^+$ such that $\alpha(w)$ contains a square, say

$$\alpha(w) = yuuz \qquad \text{with } u \neq 1.$$

Set $w = a_1 a_2 \cdots a_n$, $v_i = \alpha(a_i)$ $(a_i \in A)$. By condition (i), one has $n \geqslant 4$. Next y is a proper left factor of v_1 and z is a proper right factor of v_n since w was chosen shortest. Also yu is not a left factor of v_1 since otherwise $v_2 v_3$ is a factor of u, hence of v_1, violating condition (ii). For the same reason, uz is not a right factor of v_n. Thus there is an index j $(1 < j < n)$ and a factorization

$$v_j = st$$

such that (see Figure 2.2(i))

$$yu = v_1 \cdots v_{j-1}s, \qquad uz = tv_{j+1} \cdots v_n.$$

We may assume $s \neq 1$, since otherwise $j - 1 \neq 1$ and we can replace v_j by v_{j-1}. Next, define y' and z' by

$$v_1 = yy', \qquad v_n = z'z.$$

As mentioned before, y' and z' are nonempty. Further (see Figure 2.2(ii))

$$u = y'v_2 \cdots v_{j-1}s,$$
$$u = tv_{j+1} \cdots v_{n-1}z', \qquad (2.3.2)$$

Now, we derive a contradiction by showing that w contains a square. Consider first the case where $yt = 1$. In this case, $v_1 = y'$, $v_j = s$, whence by Eqs. (2.3.2)

$$u = v_1 v_2 \cdots v_{j-1} v_j = v_{j+1} \cdots v_{n-1}z'.$$

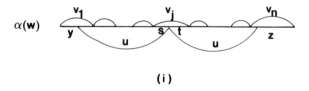

(i)

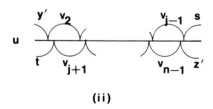

(ii)

Figure 2.2. Occurrence of uu in $\alpha(w)$: (i) localization of uu, (ii) double factorization of u.

Since $\alpha(A)$ is a prefix code, this implies $v_1 = v_{j+1}, \ldots, v_{j-1} = v_{n-1}$, $v_j = z'$, and since $v_j = z' \leqslant v_n$, we have $v_n = v_j$. Thus $w = (a_1 \cdots a_j)^2$ is a square.

Next consider the case $yt \neq 1$. Multiplying (2.3.2) by s and z, and by y and t, gives

$$suz = (sy')v_2 \cdots v_{j-1}(sz) = v_j v_{j+1} \cdots v_{n-1} v_n \qquad (2.3.3)$$

$$yut = v_1 v_2 \cdots v_{j-1} v_j = (yt) v_{j+1} \cdots v_{n-1}(z't) \qquad (2.3.4)$$

Consider Eq. (2.3.3) first. Then the claim can be applied to each of the $v_2, \ldots, v_{j-1}$. Consequently $a_2 \cdots a_{j-1}$ is a factor of $a_j a_{j+1} \cdots a_{n-1} a_n$. Since $s \neq 1$, $a_2 \cdots a_{j-1}$ is neither a left nor a right factor of $a_j, \ldots, a_n$; thus $a_2 \cdots a_{j-1}$ is a factor of $a_{j+1} \cdots a_{n-1}$ and

$$pa_2 \cdots a_{j-1}q = a_{j+1} \cdots a_{n-1} \qquad (2.3.5)$$

for some $p, q \in A^*$. Now consider Eq. (2.3.4). As before, $a_{j+1} \cdots a_{n-1}$ is a factor of $a_1 \cdots a_j$, and since neither yt nor z' is the empty word, $a_{j-1} \cdots a_{n-1}$ is a factor of $a_2 \cdots a_{j-1}$. Thus

$$\bar{p}a_{j+1} \cdots a_{n-1}\bar{q} = a_2 \cdots a_{j-1} \qquad (2.3.6)$$

for some $\bar{p}, \bar{q} \in A^*$. By (2.3.5) and (2.3.6),

$$\bar{p}pa_2 \cdots a_{j-1}q\bar{q} = \bar{p}a_{j+1} \cdots a_{n-1}\bar{q} = a_2 \cdots a_{j-1}$$

showing that $p = \bar{p} = q = \bar{q} = 1$. Thus setting

$$x = a_2 \cdots a_{j-1} = a_{j+1} \cdots a_{n-1}$$

we have

$$w = a_1 x a_j x a_n \qquad (2.3.7)$$

whence by (2.3.3) and (2.3.4)

$$st = v_j = sy', \qquad z'z = v_n = sz, \qquad yy' = v_1 = yt$$

Thus the word

$$(a_1 a_j a_n) = v_1 v_j v_n = yt\, st\, sz$$

is not square-free. By condition (i), $a_1 a_j a_n$ is not square-free. Therefore $a_1 = a_j$ or $a_j = a_n$. In view of (2.3.7), w contains a square. This yields the contradiction. ∎

Example. A tedious but finite computation shows that the morphism α: $A^* \to A^*$ with $A = \{a, b, c\}$ defined by

$$\alpha(a) = abcab, \qquad \alpha(b) = acabcb, \qquad \alpha(c) = acbcacb$$

fulfills the two conditions of Theorem 2.3.3 and therefore is a square-free morphism.

2.4. Idempotent Semigroups

Let A be an alphabet having at least three letters. Then there are infinitely many square-free words in A^*. As already mentioned in the introduction, this fact can be rephrased as follows. Let $A^* \cup 0$ be the monoid obtained by adjoining a zero to A, and consider the congruence $\approx$ generated by

$$uu \approx 0, \quad u \in A^+.$$

Each square-free word constitutes an equivalence class modulo this congruence. Consequently the quotient monoid $A^* \cup 0 / \approx$ is infinite.

There is another situation where square-free words can be used. Let $m, n \geqslant 2$ be fixed integers and consider the congruence $\equiv$ over A^* generated by

$$u^m \equiv u^n, \quad u \in A^*. \tag{2.4.1}$$

Once more, each square-free word defines an equivalence class, and thus the monoid $A^*/\equiv$ is infinite. In fact, this result also holds for a two-letter alphabet (Brzozowski, Culik II, and Gabrielian 1971).

These considerations can be placed in the framework of the classical Burnside problem (originally, the Burnside problem was formulated for groups only, but it is easy to state for semigroups also): *Is every finitely generated torsion semigroup finite?* (A torsion semigroup is a semigroup such that each element generates a finite subsemigroup.) We have just seen that the answer is negative in general, and this is due to the existence of infinitely many square-free words. For groups, the answer also is negative (see Chapter 8 in Herstein 1968). Moreover, the groups of exponent n — that is, groups where each element has exponent n — are in general infinite (see Adjan 1979). The proof uses the fact that there are infinitely many square-free words. For another result on the Burnside problem, see Chapter 7, Section 7.3.

In one special case, surprisingly, the answer is positive. Let A be an arbitrary finite alphabet, and consider the congruence $\sim$ generated by the relations

$$ww \sim w, \qquad w \in A^*. \tag{2.4.2}$$

The quotient monoid

$$M = A^* / \sim$$

is called the *free idempotent monoid* on A; indeed, any element in M is idempotent ($mm = m$), and any finitely generated idempotent monoid is easily seen to be a quotient of a free idempotent monoid.

THEOREM 2.4.1 (Green–Rees). *The free idempotent monoid on A is finite and has exactly*

$$\sum_{k=0}^{n} \binom{n}{k} \prod_{1 \leq i \leq k} (k - i + 1)^{2^i} \tag{2.4.3}$$

elements, where $n = \mathrm{Card}(A)$.

The numbers (2.4.3) are growing very rapidly. For $n = 0, 1, 2, 3, 4$, they are $1, 2, 7, 160, 332381$.

Before starting the proof, it will be interesting to note the difference between the relations (2.4.1) and (2.4.2). For the congruence defined by (2.4.1), two distinct words can be congruent only if both contain at least one pth power, for $p = \min(m, n)$. On the contrary, two distinct square-free words may be congruent for $\sim$. Indeed, the defining relations allow introduction of squares and then dropping of other ones. We give now a nontrivial illustration of this situation by verifying that $x \sim y$ with $x = bacbcabc$ and $y = bacabc$. Both x and y are square-free words, and they are also equivalent. Indeed, note first that with $u = abcaca$, we have (boldfaced factors are those to be reduced) $uy = abc\mathbf{acabacabc} \sim abc\mathbf{aca}bc \sim \mathbf{abcabc} \sim abc$ whence $x = (bacbc)abc \sim bacbcuy = vy$ for $v = bacbcu$.

Next, for $r = bcabacbcacbcbac$, we have

$$xr = ba\mathbf{cbcabcbcab}acbcacbcbac$$
$$\sim \mathbf{bacbcabacbca}cbcbac$$
$$\sim b\mathbf{acbcacbcbac} \sim ba\mathbf{cbcb}ac \sim \mathbf{bacbac} \sim bac$$

whence

$$y = bacabc \sim xrabc \sim xs$$

with $s = rabc$. Finally,

$$x \sim vy \sim vyy \sim xy \sim xxs \sim xs \sim y,$$

which proves the claim.

Proof of Theorem 2.4.1. Recall from Chapter 1 that for $w \in A^*$,

$$\mathrm{alph}(w) = \{a \in A \mid |w|_a \neq 0\}$$

It is clear that $x \sim y$ implies $\mathrm{alph}(x) = \mathrm{alph}(y)$. First we prove the following claim:

Claim (i). *If* $\mathrm{alph}(y) \subset \mathrm{alph}(x)$, *there exists* u *such that* $x \sim xyu$.

This is indeed clear if $y = 1$. Assume $|y| \geq 1$, and let $y = y'a$ with $a \in A$. By induction, there is a word u' such that $x \sim xy'u'$. Furthermore, $a \in \mathrm{alph}(x)$, whence $x = zaz'$. Thus for $u = z'y'u'$

$$xyu = z\mathbf{az'y'az'y'}u' \sim zaz'y'u' = xy'u' \sim x.$$

This proves Claim (i).

For $x \in A^+$, let x' be the shortest left factor of x such that $\mathrm{alph}(x') = \mathrm{alph}(x)$. Setting $x' = pa$ for some $p \in A^*$, $a \in A$, we have $\mathrm{alph}(p) = \mathrm{alph}(x) - \{a\}$. Symmetrically, the shortest right factor x'' of x with $\mathrm{alph}(x'') = \mathrm{alph}(x)$ has the form $x'' = bq$ for some $b \subset A$, $q \in A^*$ and $\mathrm{alph}(q) = \mathrm{alph}(x) - \{b\}$. Thus to x there is associated a quadruple (p, a, b, q). We write this fact $x \hat{=} (p, a, b, q)$, and prove:

Claim (ii). *If* $x \hat{=} (p, a, b, q)$, *then* $x \sim pabq$.

Indeed let, $x = pay = zbq$. Since $\mathrm{alph}(y) \subset \mathrm{alph}(x) = \mathrm{alph}(pa)$, there is by (i) a word u such that $pa \sim payu = xu$. Since $\mathrm{alph}(pa) \subset \mathrm{alph}(bq)$, the dual of (i) shows that there is a word v with $bq \sim vpabq = v\hat{x}$, where $\hat{x} = pabq$. This implies that

$$\hat{x} = pabq \sim xubq = xw$$

for $w = ubq$ and

$$x = zbq \sim zv\hat{x} = t\hat{x}$$

for $t = zv$. Whence

$$x \sim t\hat{x} \sim t\hat{x}\hat{x} \sim x\hat{x} \sim xxw \sim xw \sim \hat{x}.$$

This proves (ii).

In view of Claim (ii), we can show that M is finite as follows. Assume that the finiteness holds for alphabets that have fewer elements than A. If $x \hat{=} (p, a, b, q)$, then $\mathrm{Card}(\mathrm{alph}(p)) < \mathrm{Card}(A)$ and $\mathrm{Card}(\mathrm{alph}(q)) < \mathrm{Card}(A)$, thus there are only finitely many ps and qs modulo $\sim$. Since there are only finitely many letters, M itself is finite. In order to compute the number of elements in M, we prove the following equivalence.

Claim (iii). Let $x \triangleq (p, a, b, q)$ *and* $x' \triangleq (a', a', b', q')$; *then* $x \sim x'$ *iff* $p \sim p'$, $a = a'$, $b = b'$, $q \sim q'$.

Suppose first that $p \sim p'$, $a = a'$, $b = b'$, $q \sim q'$. Then $pabq \sim p'a'b'q'$ and $x \sim x'$ by (ii). Suppose now $x \sim x'$. One can assume that $x = \alpha\beta\gamma$, $x' = \alpha\beta^2\gamma$ for some words $\alpha, \beta, \gamma \in A^*$. We distinguish two cases.

Case 1. $|\alpha\beta| > |p|$. Setting $x = pay$, we have

$$\alpha\beta = pat, \qquad z = t\gamma$$

for some t in A^+. Then $x' = pat\beta\gamma$ and $\mathrm{alph}(p) = \mathrm{alph}(x) - \{a\} = \mathrm{alph}(x') - \{a\}$. Thus by definition $p' = p$ and $a' = a$.

Case 2. $|\alpha\beta| \leqslant |p|$. Setting $x = pay$, there is a word $s \in A^*$ such that

$$p = \alpha\beta s, \qquad \gamma = sag.$$

Then $x' = \alpha\beta^2 sag$ and $\mathrm{alph}(\alpha\beta^2 s) = \mathrm{alph}(\alpha\beta s) = \mathrm{alph}(x) - \{a\} = \mathrm{alph}(x') - \{a\}$. Thus by definition $p' = \alpha\beta^2 s$ whence $p' \sim p$, $a = a'$.

The relations $b = b'$, $q \sim q'$ are proved in a symmetric manner.

We now are ready to compute the number of elements in $M = A^*/\sim$. Let $\pi: A^* \to M$ be the canonical morphism and let, for $B \subset A$,

$$\overline{B} = \{x \in A^* | \mathrm{alph}(x) = B\}.$$

Then A^* is the disjoint union of the sets $\overline{B}$, $B \subset A$. Since $x \sim x'$ implies $\mathrm{alph}(x) = \mathrm{alph}(x')$, each $\overline{B}$ is a union of equivalence classes mod $\sim$, whence M is the disjoint union of the sets $\pi(\overline{B})$, $B \subset A$.

In view of Claim (iii), if $B \neq \varnothing$, there is a bijection

$$\pi(\overline{B}) \to \bigcup_{a,b \in A} \pi\big(\overline{B - \{a\}}\big) \times \{a\} \times \{b\} \times \pi\big(\overline{B - \{b\}}\big)$$

Thus if $\mathrm{Card}(B) = k \geqslant 1$, and setting $c_k = \mathrm{Card}(\pi(\overline{B}))$, we have

$$c_k = k^2 c_{k-1}^2.$$

Clearly $c_0 = 1$, whence

$$c_k = \prod_{i=1}^{k} (k - i + 1)^{2^i}.$$

Consequently, M being the disjoint union of the $\pi(\overline{B})$,

$$\mathrm{Card}\, M = \sum_{k=0}^{n} \binom{n}{k} c_k.$$

This completes the proof.   ∎

Notes

Axel Thue was the first author to investigate avoidable regularities, especially words without overlapping factors and square-free words. His two papers (Thue 1906, 1912) on this topic contain the definitions of the words **t** and **m**, and the proofs of Theorems 2.2.3 and 2.3.1 as reported here. Theorem 2.3.3 is a slight improvement, due to Bean, Ehrenfeucht, and McNulty (1979), of a result of Thue. The infinite word **t** was discovered independently by Morse (1921, 1938), the square-freeness of **m** was proved by Morse and Hedlund in 1944, Braunholtz in 1963, and Istrail in 1977. Many other papers have been written on infinite square-free words or related topics (Arson 1937; Dean 1965; Gottschalk and Hedlund 1964; Hawkins and Mientka 1956; Leech 1957; Li 1976; Pleasants 1970; Shepherdson 1958; Zech 1958; Dekking 1976; Entringer, Jackson, and Schatz 1974; Ehrenfeucht and Rozenberg 1981; Main and Lorentz 1979; Crochemore 1981). As noted by Hedlund in 1967, some of the work done later is already contained in Thue's papers, which were forgotten for a long time.

One of the problems raised in Thue's 1912 paper that has been significantly developed concerns the distance between two occurrences of a factor in a word. Indeed, an infinite word **a** is square-free iff whenever xyx is a factor of **a** with $x \neq 1$, then $y \neq 1$. Thus one may define the number

$$e_{\mathbf{a}}(x) = \min\{|y| : xyx \text{ is a factor of } \mathbf{a}\}$$

and look for lower bounds for $e_{\mathbf{a}}(x)$. Thue gives an infinite word **a** over k letters (for each $k \geqslant 3$) such that $e_{\mathbf{a}}(x) \geqslant k - 2$ for all x occurring twice in **a**. F. Dejean (1972) improves this inequality. She constructs an infinite word **a** over three letters such that

$$e_{\mathbf{a}}(x) \geqslant \tfrac{1}{3}|x|$$

for all factors x occurring twice in x. She also shows that this lower bound is optimal. Pansiot, in a forthcoming paper, handles the case of four letters. For more than four letters, the sharp value of the lower bound remains unknown.

Square-free morphisms and more generally k th-power-free morphisms are investigated in Bean, Ehrenfeucht and McNulty 1979. Characterizations of square-free morphisms are given in Berstel 1979 and Crochemore 1982. Bean et al. introduce the very interesting concept of so-called avoidable patterns, which are described as follows:

Let E and A be two alphabets. For easier understanding, E will be called the pattern alphabet, a word in E^+ is a *pattern*. Let $w = e_1 e_2 \cdots e_n$ $(e_i \in E)$ be a pattern. A word u in A^+ is a *substitution instance* of w iff there is a nonerasing morphism $\lambda: E^* \to A^*$ such that $u = \lambda(w)$. Equivalently, $u = x_1 x_2 \cdots x_n$ with $x_1, \ldots, x_n \in A^+$ and with $x_i = x_j$ whenever $e_i = e_j$. Setting for example $E = \{e\}$, $A = \{a, b, c\}$, the word $u = abcabc$ is a substitution instance of ee.

A word u in A^+ *avoids* the pattern w in E^+ iff no factor of u is a substitution instance of w. Thus for example $u \in A^+$ avoids the pattern ee iff u is square-free, and u avoids $ee'ee'e$ iff u has no overlapping factor. Given a pattern w in E^+, w is called *avoidable on A* if there exist infinitely many words u in A^+ that avoid w. The existence of infinite square-free words, and infinite words without overlapping factor can be rephrased as follows: The word ee is avoidable on a three-letter alphabet, the word $ee'ee'e$ is avoidable on a two-letter alphabet. This formulation, of course, raises the question of the structure of avoidable patterns. Among the results of the paper of Bean et al., we report the following: Let $n = \operatorname{Card} E$; then there is a finite alphabet A such that every pattern w with $|w| \geqslant 2^n$ is avoidable on A.

Another interesting extension of square-freeness is abelian square-freeness, also called strong nonrepetitivity. An abelian square is a word uu', such that u' is a rearrangement of u, that is $|u|_a = |u'|_a$ for each letter a. A word is strongly nonrepetitive if it contains no factor that is an abelian square. Calculation shows that over three letters, every word of length $\geqslant 8$ has an abelian square. On the other hand, Pleasants (1970) has shown that there is an infinite strongly nonrepetitive word over five letters. This improves considerable the previously known bound of twenty five letters given by Evdokomov in 1968. The case of four letters is still open. For related results, see Justin 1972, T. C. Brown 1971, and Dekking 1979.

Concerning idempotent semigroups, Theorem 2.4.1 is a special case of a more general result also due to Green and Rees (1952). Let $r \geqslant 1$ be an integer. Then the two following conditions are equivalent:

(i) Any finitely generated group G such that $x^r = 1$ for all x in G is finite.
(ii) Any finitely generated monoid M such that $x^{r+1} = x$ for all x in M is finite.

The case considered in Theorem 2.4.1 is $r = 1$, and in this case the group G is trivially finite. For a proof of the theorem, see Green and Rees 1952 or Lallement 1979. Note that there are integers r such that condition (i), and consequently (ii), does not hold; $r = 665$ is such an integer (see Adian 1979). Moreover, Theorem 2.4.1 was generalized by Simon (1980) who proved the result that for a finitely generated semigroup S the following three conditions are equivalent:

(i) S is finite.
(ii) S has only finitely many nonidempotent elements.
(iii) There exists an integer m such that for each sequence $(s_1, \ldots, s_m)$ in S there exist i, j ($i < j$) such that $s_i \cdots s_j$ is idempotent.

Problems

Section 2.1

2.1.1. Let P be a property of words of A^* such that $I = \{w \mid P(w)\}$ is a two-sided ideal. Each infinite word on A has a factor in I iff $A^* - I$ is finite (*Hint*: Apply Lemma 2.1.2 to (not P).) (See Justin 1972.)

Section 2.2

2.2.1. Assume that wuu is a left factor of t. Then for some $n \geqslant 0$, $|u| = 2^n$ or $|u| = 3.2^n$ and $|w|$ is a multiple of 2^n. If φ is any morphism such that $\mathbf{t} = \varphi(\mathbf{t})$, then $\varphi = \mu^m$ for some $m \geqslant 0$. (See Pansiot 1981.)

2.2.2. To each infinite word $\mathbf{a}$ over $A = \{a, b\}$, associate the formal power series

$$\sum_{i \geqslant 0} \hat{a}_i X^i$$

in the variable X and coefficients in $\mathbb{F}_2 = \mathbb{Z}/2\mathbb{Z}$ defined by $\hat{a}_i = 0$ or 1 according to whether $\mathbf{a}(i) = a$ or b. Let y and $\bar{y}$ be the formal power series associated to $\mathbf{t}$ and $\bar{\mathbf{t}}$, respectively. Then y and $\bar{y}$ are the solutions of the equation

$$(1 + X)^3 z^2 + (1 + X)^2 z + X = 0$$

in the ring of formal power series over $\mathbb{F}_2$. (See Cristol, Kamae, Mendès-France, and Rauzy 1980.)

2.2.3. To each function $f: \mathbb{N} \to \mathbb{N}$ with $f(0) = 0$ and $f(n+1) - f(n)$ positive and odd for $n \geqslant 0$, associate the infinite word on $A = \{a, b\}$ defined by

$$\mathbf{a}_f = a v_{f(0)} v_{f(1)} \cdots v_{f(n)} \cdots,$$

where $v_k = \mu^k(b)$. Each $\mathbf{a}_f$ is an infinite word without overlapping factor, the mapping $f \mapsto \mathbf{a}_f$ is injective and consequently the set of infinite words over A without overlapping factor is not denumerable. (This is a simplified version of a construction of Kakutani, as reported in Gottschalk and Hedlund 1955.)

Section 2.3

2.3.1. For each infinite word $\mathbf{a} = a_0 a_1 \cdots a_n \cdots$ on $A = \{a, b\}$, define an infinite word $\mathbf{b} = b_0 b_1 \cdots b_n \cdots$ on $B = \{a, b, c\}$ by

$$b_n = \begin{cases} a & \text{if} \quad a_n a_{n+1} = aa \quad \text{or} \quad bb \\ b & \text{if} \quad a_n a_{n+1} = ab \\ c & \text{if} \quad a_n a_{n+1} = ba \end{cases}$$

If $\mathbf{a}$ has no overlapping factor, then $\mathbf{b}$ is square-free. If $\mathbf{a} = \mathbf{t}$, then $\mathbf{b} = \mathbf{m}$ (Morse and Hedlund 1944).

2.3.2. Prove Proposition 2.3.2 (See Istrail 1977 and also Dekking 1978.)

2.3.3. Define a sequence w_n of words over $\{a, b, c\}$ by

$$w_0 = 1,$$
$$w_{n+1} = w_n a w_n b w_n c w_n \quad (n \geqslant 0).$$

Then $\mathbf{m} = \lim w_n$. (Due to Cousineau unpublished.)

*2.3.4. Define over $A = \{a, b, c\}$ two mappings π, ι by $\pi(a) = abc$, $\pi(b) = bca$, $\pi(c) = cab$ and $\iota(d) = \pi(d)\tilde{}$ $(d \in A)$. Extend π to A^* by

$$\pi(a_1 a_2 \cdots a_n) = \pi(a_1)\iota(a_2)\pi(a_3)\iota(a_4)\cdots.$$

Then $\mathbf{a} = \pi^\omega(a)$ is an infinite square-free word. (See Arson 1937; see also Yaglom and Yaglom 1967.)

2.3.5. A morphism $\alpha: A^* \to B^*$ is called kth power-free if $\alpha(w)$ is kth power-free for each kth-power-free word w. If $\alpha: A^* \to B^*$ is a square-free morphism such that

(i) no $\alpha(a)$ is a proper factor of an $\alpha(b)$ $(a, b \in A)$,

(ii) no $\alpha(a)$ has a nonempty proper left factor that is also a right factor of $\alpha(a)$,

then α is kth power-free for all $k > 1$. (See Bean, Ehrenfeucht, and McNulty 1979.)

2.3.6. The set of infinite square-free words over three letters is not denumerable (*Hint*: Use Problem 2.2.3.)

2.3.7. With the notations of Theorem 2.3.1: $\mathbf{b}$ is a square-free word such that neither aba nor $acbca$ is a factor of $\mathbf{b}$, if and only if $\delta(\mathbf{b})$ has no overlapping factor. (See Thue 1912.)

van der Waerden's Theorem

3.0. Introduction

This chapter is devoted to a study of van der Waerden's theorem, which is, according to Khinchin, one of the "pearls of number theory." This theorem illustrates a principle of unavoidable regularity: It is impossible to produce long sequences of elements taken from a finite set that do not contain subsequences possessing some regularity, in this instance arithmetic progressions of identicai elements.

During the last fifty years, van der Waerden's theorem has stimulated a good deal of research on various aspects of the result. Efforts have been made to simplify the proof while at the same time generalizing the theorem, as well as to determine certain numerical constants that occur in the statement of the theorem. This work is of an essentially combinatorial nature. More recently, results from ergodic theory have led to the discovery of new extensions of van der Waerden's theorem, and, as a result, to a topological proof.

The plan of the chapter illustrates this diversity of viewpoints. The first section, after a brief historical note, presents several different formulations of van der Waerden's theorem. The second section gives a combinatorial proof of an elegant generalization due to Grünwald. The third section, which concerns "cadences," gives an interpretation of the theorem in terms of the free monoid. In the fourth section is presented a topological proof of van der Waerden's theorem, due to Fürstenberg and Weiss. The final section is devoted to related results and problems: estimation of various numerical constants, Szemeredi's theorem, conjectures of Erdös, and so on.

3.1. Classical Formulations

Some forty years after he proved the theorem that bears his name, van der Waerden published an article (1965 and 1971) in which he describes the circumstances of the theorem's discovery. In 1926, in Hamburg, the

mathematicians E. Artin, O. Schreier, and B. van der Waerden set to work
on the following conjecture of the Dutch mathematician Baudet:

PROPOSITION 3.1.1. *If* $\mathbb{N}$ *is partitioned into two classes, one of the classes
contains arbitrarily long arithmetic progressions.*

The conjecture was extended by Artin to the case of a partition of $\mathbb{N}$ into
k classes. (By a partition of a set E into k classes we mean a family
$\mathcal{E} = \{E_1,\ldots,E_k\}$ of pairwise disjoint subsets of E whose union is E). The
generalization of Baudet's conjecture is thus:

PROPOSITION 3.1.2. *If* $\mathbb{N}$ *is partitioned into* k *classes, one of the classes
contains arbitrarily long arithmetic progressions.*

The conjecture was sharpened by Schreier and proved by van der
Waerden in the following form:

THEOREM 3.1.3. (van der Waerden's theorem). *"For all integers* $k,l \in \mathbb{N}$
there exists an integer $N(k,l) \in \mathbb{N}$ *such that if the set* $\{0,1,\ldots,N(k,l)\}$ *is
partitioned into* k *classes, one of the classes contains an arithmetic progression
of length* l.

It can be shown directly that statements 3.1.2 and 3.1.3 are equivalent. In
his account, van der Waerden proposes a diagonal method that amounts to
using the following compactness argument.

Let us fix the integers k and l. If 3.1.3 is false, then for each $n \in \mathbb{N}$ there
exists a partition $\mathcal{E}_n = \{E_{n,1},\ldots,E_{n,k}\}$ of $\{0,\ldots,n\}$ into k classes such that
no class of $\mathcal{E}_n$ contains an arithmetic progression of length l. We associate to
each $\mathcal{E}_n$ a sequence $\chi_n \in \{0,\ldots,k\}^{\mathbb{N}}$ defined by

$$\chi_n(i) = \begin{cases} j & \text{if} \quad i \in E_{n,j} \\ 0 & \text{if} \quad i > n. \end{cases}$$

Now consider the set $K = \{0,\ldots,k\}^{\mathbb{N}}$ of all sequences with values in $\{0,\ldots,k\}$
as a topological space with the product topology: K is then a compact
metric space that admits the distance function

$$d(y_1, y_2) = \inf\left\{ \frac{1}{k+1} \middle| y_1(n) = y_2(n) \text{ for all } n \text{ such that } 0 \leq n < k \right\}.$$

Thus the sequence $(\chi_n)_{n \in \mathbb{N}}$ has at least one limit point $\chi \in K$, so for each
$r \in \mathbb{N}$ there exists an integer $n(r)$ such that the first r values of $\chi_{n(r)}$ and χ
are equal. It follows that every term of χ is in $\{1,2,\ldots,k\}$. Now consider the

partition $\mathcal{F} = \{F_1, \ldots, F_k\}$ of $\mathbb{N}$ defined by

$$t \in F_j \quad \text{iff} \quad \chi(t) = j.$$

Suppose some class F_j of $\mathcal{F}$ contains an arithmetic progression of length l. Then $\chi(t_0) = \chi(t_0 + a) = \cdots = \chi(t_0 + (l-1)a) = i$ for some $t_0 \in \mathbb{N}$ and $a > 0$. Thus for some n, $\chi_n(t_0) = \chi_n(t_0 + a) = \cdots = \chi_n(t_0 + (l-1)a) = i$, consequently the progression $\{t_0, t_0 + a, \ldots t_0 + (l-1)a\}$ is contained in $E_{n,i}$, contrary to the hypothesis. So no class of $\mathcal{F}$ contains an arithmetic progression of length l, which contradicts Proposition 3.1.2. Thus 3.1.2 implies 3.1.3. Since 3.1.3 clearly implies 3.1.2 the equivalence of the two statements is proved.

3.2. A Combinatorial Proof of van der Waerden's Theorem

The present exposition follows that of Anderson (1976), who proves the following more general result, due to Grünwald (unpublished).

THEOREM 3.2.1. *Let S be a finite subset of $\mathbb{N}^d$. For each k-coloring of $\mathbb{N}^d$ there exists a positive integer a and a point v in $\mathbb{N}^d$ such that the set $aS + v$ is monochromatic. Moreover, the number a and the coordinates of the point v are bounded by a function that depends only on S and k (and not on the particular coloring used).*

(In this statement the word "k-coloring" is synonymous with "partition into k classes." Two points of $\mathbb{N}^d$ are of the same color if they belong to the same class of the partition. A *monochromatic* set is one in which all the elements are of the same color.)

Figure 3.1 illustrates the assertion 3.2.1 in the case $k = 2$ and $d = 2$. Begin by noting that Theorem 3.2.1 implies van der Waerden's theorem. Indeed,

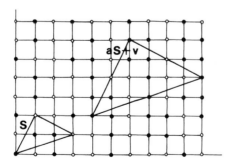

Figure 3.1. Case $d = 2, k = 2$.

since van der Waerden's theorem is equivalent to statement 3.1.2 it is enough to observe that 3.1.2 follows immediately from 3.2.1 upon taking $d = 1$ and $S = \{0, 1, \ldots, l-1\}$.

Proof (combinatoric) of Theorem 3.2.1. In the following, S denotes a finite subset of $\mathbb{N}^d$ and s an element of $\mathbb{N}^d$. K_n denotes the cube of side n consisting of those points of $\mathbb{N}^d$ all of whose coordinates are less than n. The proof consists of establishing the following statement by induction on $|S|$:

$A(S)$: for each integer $k \in \mathbb{N}$ there exists an integer $n = n(k)$ such that for every k-coloring of K_n, K_n contains a monochromatic subset of the form $aS + v$.

Theorem 3.2.1 follows from the statements $A(S)$. Indeed, the number a and the coordinates of v can be bounded by n, since $aS + v$ is contained in K_n.

Let us begin by observing that $A(S)$ is true if S is empty or consists of a single point, since any set of cardinality ≤ 1 is necessarily monochromatic. Henceforth we will suppose that S is nonempty.

To show that $A(S)$ implies $A(S \cup \{s\})$ we introduce an auxiliary statement $C(p)$, where S is fixed and $p \in \mathbb{N}$.

$C(p)$: Let $k \in \mathbb{N}$ and $s \in \mathbb{N}^d$. Then there exists an integer $n = n(p, k, s)$ such that for each k-coloring of K_n there exist positive integers $a_0, a_1, \ldots, a_p$ and a point $u \in \mathbb{N}^d$ such that the $(p+1)$ sets

$$T_q = u + \left(\sum_{0 \leq i < q} a_i \right) S + \left(\sum_{q \leq i \leq p} a_i \right) s \quad (0 \leq q \leq p)$$

are monochromatic subsets of K_n.

These "intermediate" assertions between $A(S)$ and $A(S \cup \{s\})$ are proved by induction on p.

$p = 0$: Choose n such that $s \in K_n$ and then set $u = (0, \ldots, 0)$ and $a_0 = 1$. Then $T_0 = \{s\}$ is a monochromatic set contained in K_n. *Passage from p to $p+1$*: Let $n = n(p, k, s)$ be the integer specified in the statement $C(p)$ and let $k' = k^{n^d}$. If k colors are available, then there are k' ways to color the cube K_n (since K_n contains n^d points). Thus each k-coloring of $\mathbb{N}^d$ induces a k'-coloring of $\mathbb{N}^d$: Two points u and v will have the same color in the new k'-coloring if and only if the cubes $u + K_n$ and $v + K_n$ are colored identically in the original k-coloring of $\mathbb{N}^d$. It follows from $A(S)$ that there exists an integer $n' = n'(k')$ such that for every k'-coloring of $K_{n'}$, $K_{n'}$ contains a monochromatic subset of the form $a'S + v'$.

Let $N = n + n'$, and consider a k-coloring of K_N. This can be extended, in arbitrary fashion to a k-coloring of $\mathbb{N}^d$, which induces, as described previously, a k'-coloring of $\mathbb{N}^d$. Since $N > n'$, K_N contains a monochromatic subset (with respect to the induced coloring) of the form $a'S + v'$. This means that the $|S|$ cubes $K_n + a't + v'$ (where t runs over all the points of S) are colored identically with respect to the original k-coloring. However by $C(p)$, K_n contains monochromatic sets

$$T_q = u + \left(\sum_{0 \leqslant i < q} a_i \right) S + \left(\sum_{q \leqslant i \leqslant p} a_i \right) s \quad (0 \leqslant q \leqslant p).$$

It follows, upon setting $b_0 = a'$ and $b_i = a_{i-1}$ for $1 \leqslant i \leqslant p+1$, that the sets

$$T'_q = (u + v') + \left(\sum_{0 \leqslant i < q} b_i \right) S + \left(\sum_{q \leqslant i \leqslant p+1} b_i \right) s \quad (0 \leqslant q \leqslant p+1)$$

are monochromatic. Indeed, if $1 \leqslant q \leqslant p+1$, $T'_q = T_{q-1} + a'S + v'$, which is monochromatic by construction. If $q = 0$,

$$T'_0 = (u + v') + \left(\sum_{0 \leqslant i \leqslant p+1} b_i \right) s$$

is a singleton and hence monochromatic. Thus $C(p+1)$ holds, with $n(p+1, k, s) = N$.

Let us now prove $A(S \cup \{s\})$. Fix the number of colors k and apply $C(k)$: There is an integer n such that for every k-coloring of K_n, there exist $k+1$ monochromatic sets $T_0, \ldots, T_k$. By the pigeonhole principle, two of these sets (say, T_r and T_q, where $r < q$) must be of the same color:

$$T_r = u + \left(\sum_{0 \leqslant i < r} a_i \right) S + \left(\sum_{r \leqslant i < q} a_i \right) s + \left(\sum_{q \leqslant i \leqslant k} a_i \right) s$$

$$T_q = u + \left(\sum_{0 \leqslant i < r} a_i \right) S + \left(\sum_{r \leqslant i < q} a_i \right) S + \left(\sum_{q \leqslant i \leqslant k} a_i \right) s$$

Since we supposed S to be nonempty at the beginning of the induction, S contains at least one point s_0. It follows that the set

$$T = u + \left(\sum_{0 \leqslant i < r} a_i \right) s_0 + \left(\sum_{r \leqslant i < q} a_i \right) (S \cup \{s\}) + \left(\sum_{q \leqslant i \leqslant k} a_i \right) s$$

is monochromatic and contained in K_n.

Setting

$$a = \sum_{r \leqslant i < q} a_i \quad \text{and} \quad v = u + \left(\sum_{0 \leqslant i < r} a_i \right) s_0 + \left(\sum_{q \leqslant i \leqslant k} a_i \right) s,$$

we find that $T = a(S \cup \{s\}) + v$, which establishes $A(S \cup \{s\})$ and completes the proof by induction. ∎

3.3. Application to the Free Monoid

The following interpretation of van der Waerden's theorem in words is based on the definition of *cadences*, first introduced by Gardelle and Guilbaud (1964) in a slightly different form.

Let $T = \{t_1, \ldots, t_n\}$ be a finite subset of $\mathbb{N}$ with $0 < t_1 < t_2 \cdots < t_n$ and let $u = a_1 \cdots a_r (a_1, \ldots, a_r \in A)$ be a word in A^*. T is a *cadence* of u if $t_n \leqslant r = |u|$ and if $a_{t_1} = a_{t_2} = \cdots = a_{t_n}$. The number n is called the *order* of the cadence.

Let S be a finite subset of $\mathbb{N}$. A *cadence of type S* of u is a cadence of u of the form $\alpha S + \beta$, where α and β are positive integers.

A cadence of the form $T = \alpha S + \beta$ (where $\alpha, \beta > 0$ and $S = \{0, 1, \ldots, n-1\}$) is called an arithmetic cadence with common difference α.

Example. The word *abbabbabbaab* has an arithmetic cadence of order 4 ($\alpha = 3$, $\beta = 3$, $S = \{0, 1, 2, 3\}$). The word *abbaabba* has no arithmetic cadence of order greater than 2. The set $\{1, 4, 5, 8\}$ is a cadence of this word.

Now that we have the definitions we can reformulate van der Waerden's theorem. To do this, it is enough to observe that a word u of length n over a k-letter alphabet A partitions the set $\{1, \ldots n\}$ into k classes, each class consisting of those positions in u where a particular letter occurs. We thus obtain a bijection between A^n and the set of partitions of $\{1, \ldots, n\}$ into k classes, giving the following interpretations of Theorems 3.1.3 and 3.2.1 (with $d = 1$).

PROPOSITION 3.3.1. *If A is an alphabet with k letters and n is an integer, there exists an integer $N = N(k, n)$ such that each word of length $\geqslant n$ has an arithmetic cadence of order n.*

PROPOSITION 3.3.2. *Let S be a finite subset of $\mathbb{N}$ and A be an alphabet with k letters. There exists an integer N, depending only on S and k, such that every word of length $\geqslant N$ has a cadence of type S.*

We can, moreover, show directly that these two statements are equivalent: Proposition 3.3.1 follows from 3.3.2 by setting $S = \{0, 1, \ldots, n-1\}$. Conversely, let S be a finite subset of $\mathbb{N}$ and let m be the largest element of S. By (3.3.1), there exists an integer N such that every word w in A^* of length $\geqslant N$ has an arithmetic cadence of order $m + 1$, that is, a cadence of the form $\alpha T + \beta$, where $T = \{0, 1, \ldots, m\}$, $\alpha > 0$ and $\beta > 0$. Since $S \subset T$, it follows that $\alpha S + \beta$ is also a cadence of w that establishes Proposition 3.3.2.

3.4. A Topological Proof of van der Waerden's Theorem

The source for the purely topological proof of van der Waerden's theorem to be presented here is the 1978 article of Fürstenberg and Weiss, from which this section borrows considerably. The interested reader should consult this article for the topological tools used in the proof, along with numerous extensions of the theorem. In the present exposition effort has been made as far as possible to limit the references to topology. We only suppose that the reader is familiar with the usual topological terminology: open set, closed set, compact space, continuous function, limit point, metric space, and so forth.

A topological argument has already been used in Section 3.1 to show that van der Waerden's theorem is equivalent to the following statement:

If $\mathbb{N}$ is partitioned into k classes, one of the classes contains arbitrarily long arithmetic progressions.

This is the proposition that we will now prove again by topological means.

Let $\mathcal{C} = \{C_1, C_2, \ldots, C_k\}$ be a partition of $\mathbb{N}$ into k classes and let p be a positive integer. Consider the space $E = \{1, 2, \ldots, k\}^{\mathbb{Z}}$ of all functions from $\mathbb{Z}$ into $\{1, \ldots, k\}$ with the product topology: This is a compact metrizable space that admits the distance function

$$d(u, u') = \inf\left\{\frac{1}{r+1}\,\middle|\,u(n) = u'(n) \quad \text{for all} \quad n \quad \text{such that} -r < n < r\right\}.$$

Let $v \in E$ be the function defined by

$$v(n) = \begin{cases} r & \text{if} \quad n \geqslant 0 \quad \text{and} \quad n \in C_r \\ 1 & \text{if} \quad n < 0 \end{cases}$$

We will show that there exist positive integers m and n such that

$$v(m) = v(m+n) = \cdots = v(m+pn) \tag{3.4.1}$$

that is, that $C_{v(m)}$ contains an arithmetic progression of $(p+1)$ terms. Let $S: E \to E$ be the shift operator defined by $(Su)(n) = u(n+1)$ for all $n \in \mathbb{Z}$. S is a homeomorphism from E onto itself. Let X be the set of limit points of the sequence $(S^n v)_{n \in \mathbb{N}}$. X is nonempty because E is compact, and X is closed in E and therefore is itself compact. We will show that there is a closed nonempty subset K of X that is stable under S (that is $S(K) = K$) and minimal with respect to these properties. This follows from Zorn's lemma. Indeed, the set $\mathfrak{S}$ of closed sets that satisfy these conditions (which is nonempty, since $X \in \mathfrak{S}$) is ordered by inclusion. Let $\mathfrak{F} = (F_i)_{i \in I}$ be a totally ordered family of elements of $\mathfrak{S}$. Then $F = \cap_{i \in I} F_i$ is stable under S and is contained in X. Now since $\mathfrak{F}$ is totally ordered, every finite subfamily of $\mathfrak{F}$ has a nonempty intersection. It follows from the compactness of X that F is nonempty. Thus F is a lower bound of $\mathfrak{F}$ and Zorn's lemma can now be applied. The set K was introduced for the purposes of the following proposition, which is the key to the proof:

PROPOSITION 3.4.1. *For each $\varepsilon > 0$ there exists an element z of K and an integer $n > 0$ such that $d(S^n z, z) < \varepsilon$, $d(S^{2n} z, z) < \varepsilon, \ldots, d(S^{pn} z, z) < \varepsilon$.*

This proposition will be proved by induction on p. For $p = 1$, it suffices to observe that if $x \in K$, then the sequence $(S^n x)_{n \in \mathbb{N}}$ has a limit point in K: thus for each $\varepsilon > 0$ there exist integers $i < j$ such that $d(S^j x, S^i x) < \varepsilon$. We find, upon setting $z = S^i x$ and $n = j - i$, that $d(S^n z, z) < \varepsilon$, and the result follows.

Suppose now that the proposition is true for all positive integers up through $p - 1$. We will need two intermediate lemmas.

LEMMA 3.4.2. *For each $\varepsilon > 0$ there exists a finite set of integers $k_1, \ldots, k_N$ such that for all $a, b \in K$, $\min_{1 \leqslant i \leqslant N} d(S^{k_i} a, b) < \varepsilon$.*

Proof. Here is where we use the fact that K is minimal. Recall that the only closed sets—and, by complementation, the only open sets—of K that are stable under S are $\varnothing$ and K. Thus if ω is a nonempty open subset of K, $\cup_{n \in \mathbb{Z}} S^n \omega$ is a nonempty open subset of K stable under S, and hence the family $(S^n \omega)_{n \in \mathbb{Z}}$ covers K. Since K is compact, a finite subfamily covers K. Let $\{\omega_1, \ldots, \omega_n\}$ be a finite covering of K by open sets of diameter $< \varepsilon$. Then, for $i = 1, \ldots, n$, there exists a finite family $\{S^{n_{i,j}} \omega_i\}_{1 \leqslant j \leqslant r_i}$ that covers K. Let $a, b \in K$. Then $b \in \omega_i$ for some $i \in \{1, \ldots, n\}$. Since $\{S^{n_{i,j}} \omega_i\}_{1 \leqslant j \leqslant r_i}$ covers K, $a \in S^{n_{i,j}} \omega_i$ for some $j \in \{1, \ldots, r_i\}$. Therefore $S^{-n_{i,j}} a \in \omega_i$ and thus $d(S^{-n_{i,j}} a, b) < \varepsilon$. It follows that

$$\min_{1 \leqslant i \leqslant n} \min_{1 \leqslant j \leqslant r_i} d(S^{-n_{i,j}} a, b) < \varepsilon,$$

which proves the lemma.   ∎

LEMMA 3.4.3. *For all $\varepsilon > 0$ and for all $a \in K$ there exist $b \in K$ and $n > 0$ such that $d(S^n b, a) < \varepsilon, \ldots, d(S^{pn} b, a) < \varepsilon$.*

Proof. By Lemma 3.4.2 there exist integers $k_1, \ldots, k_N$ such that for all $a, b \in K$, $\min_{1 \leqslant i \leqslant N} d(S^{k_i} a, b) < \varepsilon / 2$.

Since each S^{k_i} is uniformly continuous on K, there exists a positive real number η such that $d(a, a') < \eta$ implies $d(S^{k_i}a, S^{k_i}a') < \varepsilon/2$ for $i = 1, 2, \ldots, N$.

It follows from the inductive hypothesis that there exist $a_0 \in K$ and $n > 0$ such that

$$d(S^n a_0, a_0) < \eta, \ldots, d(S^{(p-1)n}a_0, a_0) < \eta$$

Setting $b_0 = S^{-n}a_0$ we obtain

$$d(S^n b_0, a_0) < \eta, \ldots, d(S^{pn}b_0, a_0) < \eta,$$

from which it follows that

$$d(S^{n+k_i}b_0, S^{k_i}a_0) < \varepsilon/2, \ldots, d(S^{pn+k_i}b_0, S^{k_i}a_0) < \varepsilon/2$$
$$\text{for} \quad i = 1, \ldots, N. \quad (3.4.2)$$

For each $a \in K$ there is an index j such that $d(S^{k_j}a_0, a) < \varepsilon/2$. Finally, setting $b = S^{k_j}b_0$ yields

$$d(S^n b, a) < \varepsilon, \ldots, d(S^{pn}b, a) < \varepsilon, \quad (3.4.3)$$

from which Lemma 3.4.3 follows.   ∎

The proof of the proposition can now be completed. Let a_0 be a point in K. We will construct by induction a sequence $a_1, \ldots, a_r$ of points in K, a sequence $n_1, \ldots, n_r$ of positive integers and a sequence $\varepsilon_1, \ldots, \varepsilon_r$ of real numbers $< \varepsilon/2$, such that for all $i > 0$,

$$d(S^{n_i}a_i, a_{i-1}) < \varepsilon_i/2, \ldots, d(S^{pn_i}a_i, a_{i-1}) < \varepsilon_i/2. \quad (3.4.4)$$

Set $\varepsilon_1 = \varepsilon/2$. By Lemma 3.4.2, there exist $a_1 \epsilon K$ and $n_1 > 0$ such that Eq. (3.4.4) is satisfied (with $i = 1$). By induction, assume that the construction is done for $1 \leq i \leq r$ and choice $\varepsilon_{r+1} < \varepsilon/2$ such that $d(a, a') < \varepsilon_{r+1}$ implies

$$d(S^{n_r}a, S^{n_r}a') < \varepsilon_r/2 \cdots d(S^{pn_r}a, S^{pn_r}a') < \varepsilon_r/2. \quad (3.4.5)$$

By Lemma 4.4.3, there exist $a_{r+1} \in K$ and $n_{r+1} > 0$ such that Eq. (3.4.4) is satisfied with $i = r + 1$.

Actually the following result holds for all i, j such that $0 < i \leq j$

$$d(S^{n_j + \cdots + n_i}a_j, a_{i-1}) < \varepsilon_i \cdots d(S^{p(n_j + \cdots + n_i)}a_j, a_{i-1}) < \varepsilon_i. \quad (3.4.6)$$

We prove (3.4.6) by induction on $j - i$. For $j - i = 0$, the result follows from

Eq. (3.4.4). Assume now $j - i > 0$. We have by induction

$$d\left(S^{n_j + \cdots + n_{i+1}}a_j, a_i\right) < \varepsilon_{i+1} \cdots d\left(S^{p(n_j + \cdots n_{i+1})}a_j, a_i\right) < \varepsilon_{i+1}. \quad (3.4.7)$$

Hence by Eq. (3.4.5)

$$d\left(S^{n_j + \cdots + n_i}a_j, S^{n_i}a_i\right) < \varepsilon_i / 2 \cdots d\left(S^{p(n_j + \cdots + n_i)}a_j, S^{pn_i}a_i\right) < \varepsilon_i / 2 \tag{3.4.8}$$

Therefore Eq. (3.4.6) follows from (3.4.4) and (3.4.8).

Since K is compact, there exists a pair $i < j$ such that $d(a_i, a_j) < \varepsilon/2$. It follows from (3.4.6) that

$$d\left(S^n a_j, a_j\right) < \varepsilon \cdots d\left(S^{pn}a_j, a_j\right) < \varepsilon \tag{3.4.9}$$

with $n = n_{i+1} + \cdots + n_j$. This proves Proposition 3.4.1.

Proposition 3.1.2 now follows easily. Indeed, Proposition 3.4.1 implies that there exists an element z of K and $n > 0$ such that $d(S^n z, z) < \frac{1}{2}, \ldots, d(S^{pn}z, z) < \frac{1}{2}$.

By the definition of the distance function, $z(0) = S^n z(0) = \cdots = S^{pn}z(0)$, or, equivalently,

$$z(0) = z(n) = \cdots = z(pn). \tag{3.4.10}$$

On the other hand, since $z \in K, z$ is a limit point of the sequence $(S^n v)_{n \in \mathbb{N}}$. Thus there exists an integer $m > 0$ such that $d(S^m v, z) < 1/(pn + 1)$. Hence

$$z(i) = (S^m v)(i) = v(m + i) \quad \text{for} \quad 0 \leqslant i \leqslant pn \tag{3.4.11}$$

The Proposition 3.4.1 now follows at once from Eqs. (3.4.10) and (3.4.11), completing the proof of Proposition 3.1.2.

3.5. Further Results and Problems

The principal unsolved problem concerns the values of the numerical constants that occur in the statement of van der Waerden's theorem. Let us denote by $N(k, l)$ the smallest integer for which every partition of $\{1, \ldots, N(k, l)\}$ into k classes contains at least one arithmetic progression of length l. Upper bounds on $N(k, l)$ obtained directly from the combinatoric proof of the theorem are astronomical, as they involve pileups of exponents. (More precisely for $k \geqslant 2$ fixed, these upper bounds are not even primitive recursive functions of l). Moreover, no "reasonable" upper bounds are known. The first few values of $N(k, l)$ are known:

l \ k	1	2	3	4	5
1	1	1	1	1	1
2	2	3	4	5	6
3	3	9	27		
4	4	35			
5	5	178			

Because upper bounds appear to be out of reach, most of the work has been aimed at finding lower bounds for $N(k, l)$.

In 1952 Erdös and Rado proved the following result:

$$N(k, l) \geqslant \left(2(l-1)k^{l-1}\right)^{1/2} \tag{3.5.1}$$

A combinatorial proof of (3.5.1) will be given shortly. This bound was improved in 1960 by Moser, who used a constructive method.

$$N(k, l) \geqslant lk^{c \log k}, \tag{3.5.2}$$

where c is a constant.

The other known results concern $N(2, l)$. In 1962 Schmidt used a probabilistic method to prove

$$N(2, l) \geqslant 2^{l - c(l \log l)^{1/2}}, \tag{3.5.3}$$

where c is a constant.

The best lower bound known was found by Berlekamp (1968), by means of a constructive method using the finite fields $GF(2^l)$:

If l is a prime and $l \geqslant 5$, then $N(2, l+1) > l2^l$. $\qquad$ (3.5.4)

In one of his survey papers, Erdös (1977) also mentions the following lower bound, which is valid for every l.

$$N(2, l) \geqslant c2^l, \tag{3.5.5}$$

where c is a constant. It appears that these bounds are still quite rough. Indeed, Erdös has conjectured that $N(2, l)^{1/l} \to \infty$, but, so far as we know, this has not been proved.

To prove the lower bound, Eq. (3.5.1): Set $N = N(k, l+1)$. We will begin by counting the arithmetic progressions of length $l+1$ in $\{1,\ldots,N\}$. First of all, the common difference r of such a progression can vary between 1 and $\lfloor (N-1)/l \rfloor$. Second, the first term of the progression lies between 1 and $N - rl$. It follows that the number of progressions is

$$M = \sum_{1 \leqslant r \leqslant \left\lfloor \frac{N-1}{l} \right\rfloor} (N - rl) = \frac{1}{2} \left\lfloor \frac{N-1}{l} \right\rfloor \left(2N - l - l \left\lfloor \frac{N-1}{l} \right\rfloor \right)$$

$$< \frac{1}{2} \frac{N-1}{l} (2N - (N-1))$$

and thus

$$M < \frac{N^2 - 1}{2l} < \frac{N^2}{2l}$$

On the other hand, the number of partitions of $\{1,\ldots,N\}$ into k classes is k^N. Let $1 \leqslant a_1 < a_2 \cdots < a_{l+1} \leqslant N$ be an arithmetic progression with $(l+1)$ terms. The number of partitions of $\{1,\ldots,N\}$ into k classes such that one of the classes contains this progression is k^{N-l} (indeed, the class of each element of $\{1,\ldots,N\}\backslash\{a_2,\ldots,a_{l+1}\}$ can be chosen arbitrarily, then the elements $a_2,\ldots,a_{l+1}$ must go into the class containing a_1).

It follows that the number of partitions of $\{1,\ldots,N\}$ into k classes such that at least one of the classes contains an arithmetic progression of length $l+1$ is at most Mk^{N-l}. However, by the definition of N every partition of $\{1,\ldots,N\}$ into k classes satisfies this property. Thus $k^N \leqslant Mk^{N-l}$, from which it follows that $k^l \leqslant M < N^2/2l$ and $N = N(k, l+1) > (2lk^l)^{1/2}$.

More than forty years ago, Erdös and Turan (1936) introduced another numerical constant connected with van der Waerden's theorem: Let us denote by $r_l(n)$ the smallest integer such that every subset of $\{1,\ldots,n\}$ with $r_l(n)$ elements contains an arithmetic progression of length l. For example the reader can check that $r_4(10) = 9$ as $\{1,2,3,5,6,8,9,10\}$ contains no arithmetic progression of length 4, but every subset of $\{1,\ldots,10\}$ with nine elements contains such a progression. This example is taken from an article by Wagstaff (1979) that contains a table of the first few values of $r_l(n) - 1$.

As Szemeredi (1975) has pointed out, there is good reason to study these numbers. Estimation of $r_l(n)$ is in itself interesting, and moreover, it can eventually lead to upper bounds for $N(k, l)$. Indeed, it is easy to see that if $r_l(n)/n < 1/k$ for some integer n, then $N(k, l) < n/k$. Finally, estimation of $r_l(n)$ is related to an old conjecture in number theory that asserts the existence of arbitrarily long arithmetic progressions of primes. Let $\pi(n)$ denote the number of primes $\leqslant n$; to prove the conjecture it would suffice to show that for each l the inequality $r_l(n) < \pi(n)$ holds for at least one value of n.

Here is a short historical summary of estimations of $r_l(n)$: The triangle inequality

$$r_l(m+n) \leqslant r_l(m) + r_l(n), \qquad (3.5.6)$$

proved by Erdös and Turan in 1936, leads to a fairly easy proof (see Problem 3.5.2) that

$$\lim_{n \to \infty} \frac{r_l(n)}{n} = \inf_{n>0} \frac{r_l(n)}{n} = c_l \qquad (3.5.7)$$

(cf. Behrend 1938). Erdös and Turan (1936) showed that

$$c_3 \leqslant 3/8 \qquad (3.5.8)$$

and conjectured that $c_l = 0$ for all l. In 1938 Behrend showed that if the conjecture is false then $\lim_{l \to \infty} c_l = 1$.

In 1942 Salem and Spencer proved the inequality

$$n^{l - \frac{c}{\log\log n}} < r_3(n) \qquad (3.5.9)$$

for sufficiently large n. This lower bound was improved by Behrend (1946):

$$ne^{-c\sqrt{\log n}} < r_3(n). \qquad (3.5.10)$$

Furthermore, Moser (1953) found an infinite sequence that contains no arithmetic progression of length 3 and that makes it possible to show that Eq. (3.5.10) holds for all $n > 0$. Behrend's result was generalized by Rankin (1960), who found the following lower bounds

$$ne^{-c(\log n)^{b_l}} < r_l(n) \quad \text{where } b_l = \left\lceil \frac{\log l}{\log 2} \right\rceil. \qquad (3.5.11)$$

The best upper bound for $r_3(n)$ now known is due to Roth (1952):

$$r_3(n) < \frac{cn}{\log\log n}. \qquad (3.5.12)$$

This inequality implies, of course, that $c_3 = 0$. The equality

$$c_4 = 0 \qquad (3.5.13)$$

was proved for the first time by Szemeredi (1969) by purely combinatoric methods employing van der Waerden's theorem. A bit later Roth (1970, 1972) gave an analytic proof which did not use van der Waerden's theorem.

Furthermore, Szemeredi (1975) said that Roth's method probably gives an upper bound of the following form:

$$r_4(n) < \frac{n}{\log^{(k)}n},$$

where k is a sufficiently large (but fixed) integer and where $\log^{(k)}n$ designates the k-fold iterated logarithm. Finally, Szemerédi (1975) proved the conjecture of Erdös and Turan:

$$c_l = 0 \quad \text{for all } l \tag{3.5.14}$$

and collected the \$1,000 reward offered by Erdös. Recently, Fürstenberg (1977), using ergodic theory, gave another proof of this result. (In fact Fürstenberg proves a much more general result.) Nevertheless, the problem of precise estimation of $r_l(n)$ remains open.

In a more direct formulation of Szemeredi's theorem we say that a subset S of $\mathbb{N}$ has density d if

$$\lim_{n \to \infty} \frac{S \cap \{1,\ldots,n\}}{n} = d.$$

Szemeredi's theorem can be stated as follows: *If S is a subset of $\mathbb{N}$ with density > 0, then S contains arbitrarily long arithmetic progressions.*

Erdös has offered \$3,000 for the resolution of the following conjecture, which generalizes the preceding statement: *If S is a subset of $\mathbb{N}\setminus\{0\}$ such that*

$$\sum_{s \in S} \frac{1}{s} = \infty,$$

then S contains arbitrarily long arithmetic progressions.

This conjecture, if true, would prove the conjecture on arithmetic progressions of prime numbers (take S to be the set of primes). Let us point out, with regard to prime numbers, that there exist infinitely many arithmetic progressions consisting of three primes (Chowla 1944), and that there exists an arithmetic progression consisting of seventeen primes (Weintraub 1977). However, little is known on Erdös's conjecture; see Gerver (1977).

Notes

Van der Waerden's original proof can be found in van der Waerden (1927). See also van der Waerden's personal account (1965, 1971). An exposition by Khinchin (1952) and a short proof by Witt (1951), Graham

and Rothschild (1974), Deuber (1982) are also available. The generalization
by Anderson (1976) given in the text follows essentially the arguments of
Witt and Graham–Rothschild.

Erdös's successive survey papers (1963, 1965, 1974, 1977, 1979)—with
Spencer in 1974 and with Graham in 1979—give a good idea of the
advances on van der Waerden's theorem and related topics. In particular the
reader is referred to the paper by Erdös and Graham (1979) for a number of
interesting questions not discussed here and for further references. See also
the recent book by Graham, Rothschild and Spencer (1980).

The ergodic proof of Szemeredi's theorem is due to Fürstenberg (1977).
See also Thouvenot (1978). An analogue of Szemeredi's theorem in higher
dimensions has been proved by Fürstenberg and Katznelson (1978). The
topological proof given in the text follows Fürstenberg and Weiss (1978),
where other extensions are discussed. Girard (1982) has shown that, after a
slight modification, the topological proof leads to an upper bound for
$N(k, l)$. For an application of van der Waerden's theorem to number
theory, see Shapiro and Spencer (1972).

Problems

Section 3.1

3.1.1. Let $a = (a_i)_{i \in \mathbb{N}}$ be a strictly increasing sequence of positive integers
such that there exists an integer M with $a_{i+1} - a_i \leqslant M$ for all $i \in \mathbb{N}$.
Show (without Szemerédi's theorem) that a contains arbitrarily long
arithmetic progressions. (*Hint*: Consider the classes

$$C_k = \{n \in N \,|\, \exists i \in \mathbb{N}, a_i + k = n < a_{i+1}\}$$

for $0 \leqslant k < M - 1$ and observe that $C_k \subset C_0 + k$.) (See Brown 1969,
Rabung 1970.)

3.1.2. Show that if $\mathbb{N}$ is partitioned into two classes, either one class
contains arbitrarily long sequences of consecutive integers or both
classes contain arbitrarily long arithmetic progression. (Observe that
if the first condition does not hold, then there exists an integer M
such that every interval of length M meets both classes. Now apply
Problem 3.1.1.)

Section 3.3

3.3.1. Let u be an infinite word and T a subset of $\mathbb{N}\setminus\{0\}$. We say that T is a
cadence for u if all the letters whose position in u belong to T are
identical. The definitions of cadence of type S and arithmetic cadence
are the same as in the finite case. Show that the infinite word
$aba^2b^2a^3b^3 \cdots a^nb^n \cdots$ contains no arithmetic cadence of infinite order.

3.3.2. Let u be an infinite word whose ith letter is a or b, depending on whether the first occurrence of 1 (reading from right to left) in the binary representation of $i!$ is an even or odd position. (Formally

$$u[i] = \begin{cases} a & \text{if} \quad i! = n2^{2t} & \text{with} \quad n \quad \text{odd} \\ b & \text{if} \quad i! = n2^{2t+1} & \text{with} \quad n \quad \text{odd} \end{cases}.$$

Show that for all $d > 0$ there exists an integer $n(d)$ such that u contains no arithmetic cadence with common difference d and order $\geqslant n(d)$. (Justin, unpublished; see another method in T. C. Brown 1981.)

Section 3.5

3.5.1. Show that if $\mathbb{N}$ is partitioned into k classes, one of the classes contains arbitrarily long geometric progressions (*Hint*: Given a partition $\mathcal{E} = \{E_1, \ldots, E_k\}$ of $\mathbb{N}$, consider the partition $\mathcal{E}' = \{E_1', \ldots, E_k'\}$ defined by $n \in E_i'$ if and only if $2^n \in E_i$. Then apply van der Waerden's theorem.)

3.5.2. Show that if a sequence $(u_n)_{n \geqslant 0}$ of positive real numbers satisfies the triangle inequality $u_{r+s} \leqslant u_r + u_s$ for all $r, s \geqslant 0$, then $\lim_{n \to \infty} u_n / n = \inf_{n \geqslant 0} u_n / n$. (*Hint*: Show that for $n = n_0 q + r$ where $0 \leqslant r < n_0$, $a_n / n \leqslant a_{n_0} / n_0 + a_r / n$).

CHAPTER 4

Repetitive Mappings and Morphisms

4.0. Introduction

This chapter is devoted to the study of a special type of unavoidable regularities. We consider a mapping $\varphi: A^+ \to E$ from A^+ to a set E, and we search in a word w for factors of the type $w_1 w_2 \cdots w_n$, with $\varphi(w_1) = \varphi(w_2) = \cdots = \varphi(w_n)$. The mapping is called *repetitive* when such a factor appears in each sufficiently long word. This is related both to square-free words (Chapter 2), by considering the identity mapping, and to van der Waerden's theorem (Chapter 3), as will be shown later on.

It will first be shown that any mapping from A^+ to a finite set is repetitive (Theorem 4.1.1). After a direct proof of this fact, it will be shown how the result can also be deduced from Ramsey's theorem (which is stated without proof).

Investigated also is the special case where φ is a morphism from A^+ to a semigroup S. First it is proved that a morphism to the semigroup of positive integers is repetitive when the alphabet is finite (Theorem 4.2.1). Then it is proved that a morphism to a finite semigroup is *uniformly repetitive*, in the sense that the words $w_1, w_2, \ldots, w_n$ in the foregoing definition can be chosen of equal length (Theorem 4.2.2). This is, as will be shown, a generalization of van der Waerden's theorem. Finally, the chapter mentions a number of extensions and other results.

4.1. Repetitive Mappings

The notations and definitions that will be used in what follows are presented in this section. For each word $w \in A^+$ of length n we shall denote by

$$w(i), \quad (1 \leq i \leq n),$$

the ith letter of w, and

$$w(i, j) = w(i) \cdots w(j) \quad (1 \leqslant i \leqslant j \leqslant n).$$

Given a mapping

$$\varphi \colon A^+ \to E$$

from A^+ to a set E, we say that a word w is a *kth-power modulo* φ ($k \geqslant 1$) if

$$w = w_1 w_2 \cdots w_k$$

with $w_i \in A^+$ and

$$\varphi(w_1) = \varphi(w_2) = \cdots = \varphi(w_k).$$

A kth-power modulo φ is said to be *uniform* if we have in addition

$$|w_1| = |w_2| = \cdots = |w_k|.$$

The words $w_1, w_2, \ldots, w_k$ are said to be *components* of the kth-power modulo φ.

A word $w \in A^+$ is said to contain a kth-power modulo φ if it has a factor that is a kth-power modulo φ.

A mapping

$$\varphi \colon A^+ \to E$$

from A^+ to a set E is *repetitive* (*uniformly repetitive*) if for each integer k, there exists an integer l, such that each word $w \in A^+$ of length l contains a kth-power modulo φ (uniform kth-power modulo φ).

THEOREM 4.1.1. *A mapping* $\varphi \colon A^+ \to E$ *from* A^+ *to a finite set is repetitive.*

Proof. Suppose that φ is a surjective mapping from A^+ onto $E = \{1, 2, \ldots, n\}$ and for each i, $1 \leqslant i \leqslant n$, write

$$B_i = \varphi^{-1}(i).$$

Define a mapping τ from A^* to N^n by

$$\tau(u) = (i_1, i_2, \ldots, i_n)$$

where

$$i_j = \max\{l \in N \mid u \in A^* B_j^l\}.$$

If $u < v$ —that is, u is a proper left factor of v —then $\tau(u) \neq \tau(v)$. In fact, let $v = uz$, with $z \in A^+$ and $j = \varphi(z)$. If $u \in A^*B_j^l$, then $v \in A^*B_j^{l+1}$, hence the jth component of $\tau(u)$ is strictly less than that of $\tau(v)$.

Now let $k \geqslant 1$ be an integer and $w \in A^+$ be of length k^n. The images by τ of the $k^n + 1$ left factors of w, are all distinct. Therefore, there exists at least one left factor u of w and an index i, $1 \leqslant i \leqslant n$, such that

$$u \in A^*B_i^k.$$

In fact, if no such pair (u, i) exists, then

$$\tau(u) \in \{0, 1, \ldots, k-1\}^n$$

for each left factor u of w, and the number of possible values of $\tau(u)$ is at most k^n, a contradiction.

This proves that w contains a kth-power modulo φ. ∎

Example 4.1.2. Let $A = \{a_1, a_2, \ldots, a_n\}$, and let

$$\varphi: \quad A^+ \to \{1, 2, \ldots, n\}$$

be defined by

$$\varphi(w) = \max\{i \mid a_i \in \text{alph}(w)\}$$

for every $w \in A^*$.

Consider the words $w_1, w_2, \ldots, w_n$ defined by

$$w_1 = a_1^{k-1}$$

$$w_i = (w_{i-1}a_i)^{k-1}w_{i-1} \quad (2 \leqslant i \leqslant n).$$

Then w_n is of length $k^n - 1$ and does not contain any kth-power modulo φ. This example shows that the bound k^n given by the proof of Theorem 4.1.1 is optimal.

For sufficiently long words, a much stronger property of factorization can be proved as a consequence of Ramsey's theorem. This theorem is stated here without proof: see Ramsey (1932), Graham and Rothschild in Rota (1978) or Graham, Rothschild and Spencer (1980).

Let $\mathcal{P}_k(X) = \{Y \subseteq X \mid \text{card}(Y) = k\}$, the set of k-subsets of the set X.

THEOREM 4.1.3 (Ramsey). *Let r, k, n be positive integers with $k \geqslant r \geqslant 1$. There exists an integer $R(r, k, n)$ such that for each set X of cardinality*

$R(r, k, n)$ and each partition $Y_1, Y_2, \cdots, Y_n$ of $\mathcal{P}_r(X)$ in n blocks, there exists a k-subset Y of X and a block Y_i such that $\mathcal{P}_r(Y) \subseteq Y_i$.

As a consequence, we have the following:

THEOREM 4.1.4. *Let* $\varphi: A^+ \to E$ *be a mapping from* A^+ *to a set* E *with* $\operatorname{card}(E) = n$. *For each* $k \geqslant 1$, *each word* $w \in A^+$ *of length* $R(2, k+1, n)$ *contains a factor* $w_1 w_2 \cdots w_k$, *with* $w_i \in A^+$ *and*

$$\varphi(w_i \cdots w_{i'}) = \varphi(w_j \cdots w_{j'})$$

for all pairs (i, i'), (j, j') $(1 \leqslant i \leqslant i' \leqslant k$ *and* $1 \leqslant j \leqslant j' \leqslant k)$.

Proof. Let $w \in A^+$ be of length $l = R(2, k+1, n)$. We define an equivalence on the set $\mathcal{P}_2(\{1, 2, \ldots, l+1\})$ by

$$\{i, i'\} \equiv \{j, j'\} \quad \text{iff} \quad \varphi(w(i, i'-1)) = \varphi(w(j, j'-1)),$$

for $1 \leqslant i < i' \leqslant l+1$ and $1 \leqslant j < j' \leqslant l+1$. By Ramsey's theorem, there exists a $(k+1)$-subset Y of $\{1, 2, \ldots, l+1\}$ such that all the elements of $\mathcal{P}_2(Y)$ are equivalent.

Let $Y = \{i_1, i_2, \ldots, i_{k+1}\}$, with $i_1 < i_2 < \cdots < i_{k+1}$, and

$$w_j = w(i_j, i_{j+1} - 1)$$

for $1 \leqslant j \leqslant k$. The factor $w_1 w_2 \cdots w_k$ of w satisfies the property. ∎

A word $w_1 w_2 \cdots w_k$ satisfying the property stated in Theorem 4.1.4 is of course a kth-power modulo φ with components w_i. But this property is actually much stronger. For instance, for $k = 2$ we have

$$\varphi(w_1) = \varphi(w_2) = \varphi(w_1 w_2),$$

whereas we have only $\varphi(w_1) = \varphi(w_2)$ for a square modulo φ.

4.2.   Repetitive Morphisms

In this section are investigated the repetitive morphisms

$$\varphi: A^+ \to S$$

from A^+ to a semigroup S.

When S is finite, φ is repetitive by Theorem 4.1.1. We note also that the mapping φ of the example following it is actually a morphism of semigroups. Now, applying van der Waerden's theorem, we prove the following result, where $\mathbb{P}$ denotes the additive semigroup of positive integers:

THEOREM 4.2.1. *If A is finite, any morphism $\varphi: A^+ \to \mathbb{P}$ from A^+ to $\mathbb{P}$ is repetitive.*

Proof. Let $m = \max\{\varphi(a)|a \in A\}$ and $B = \{b_i|1 \leqslant i \leqslant m\}$, an alphabet with m elements. Define a morphism $\xi: A^+ \to B^+$ by

$$\xi(a) = b_l \cdots b_2 b_1$$

for $a \in A$, with $l = \varphi(a)$.

Let $k \geqslant 1$ and $w \in A^+$. When applied to the word $\xi(w)$, van der Waerden's theorem shows that there exists an integer n such that, for $|w| = n$ (and therefore $|\xi(w)| \geqslant n$), there exists an arithmetic progression $j_1, \ldots, j_k, j_{k+1}$ of rate $r \geqslant 1$, and an integer p, such that for each i, $1 \leqslant i \leqslant k+1$, the j_ith letter of $\xi(w)$ is b_p.

For each i, $1 \leqslant i \leqslant k+1$, let u_i be the shortest left factor of w such that

$$\varphi(u_i) \geqslant j_i.$$

Then

$$\varphi(u_i) = j_i + p - 1$$

since $\xi(u_i) = v_i b_p \cdots b_1$ with $|v_i| = j_i - 1$.

Now let w_i, $1 \leqslant i \leqslant k$, be defined by

$$u_{i+1} = u_i w_i.$$

Then

$$\varphi(w_i) = \varphi(u_{i+1}) - \varphi(u_i) = j_{i+1} - j_i = r.$$

Clearly $w_1 w_2 \cdots w_k$ is a kth-power modulo φ.   ■

A refinement of Theorem 4.1.1 can be proved for morphisms. It concerns uniform repetitivity and may be viewed as a generalization of van der Waerden's theorem.

THEOREM 4.2.2. *A morphism $\varphi: A^+ \to S$ from A^+ to a finite semigroup S is uniformly repetitive.*

Proof. Let $n = \mathrm{Card}(S)$. We use an induction on n. For $n = 1$, the theorem is trivial. Suppose it is true for all integer values smaller than n.

Consider the left and right zeroes of the semigroup S. If a semigroup has more than one left (right) zero, it cannot have any right (left) zero. This is easy to see.

Suppose that S has at most one left zero (the opposite case is symmetrical). Let $k \geqslant 1$ and $w \in A^{+}$. Choose the length of w to be a multiple of k, say $|w| = kl$, and denote by $w_1, w_2, \ldots, w_k$, the consecutive factors of length l of w.

If $\varphi(w_i)$ is equal to the left zero of S for each i, $1 \leqslant i \leqslant k$, then the theorem is proved. Otherwise, w has a factor of length l, say u, such that $\varphi(u)$ is not a left zero.

Let $v_i = u(1, i)$ be the left factor of u of length i, $1 \leqslant i \leqslant l$. Let l be sufficiently large and consider the sequence

$$(\varphi(v_1), \varphi(v_2), \cdots, \varphi(v_l)).$$

Van der Waerden's theorem shows that, for a positive integer p, to be chosen later, there exists an element $s \in S$ and a pair (j, r) with $1 \leqslant j \leqslant l$ and $r \geqslant 1$, such that

$$\varphi(v_j) = \varphi(v_{j+r}) = \cdots = \varphi(v_{j+pr}) = s.$$

Let

$$v_{j+ir} = v_{j+(i-1)r} y_i$$

for $1 \leqslant i \leqslant p$. Then $s\varphi(y_i) = s$. Let $Y = \{y_1, y_2, \ldots, y_p\}$. The subsemigroup $\varphi(Y^{+})$ cannot be equal to all of S since its elements t satisfy $st = s$ and s is not a left zero of S, (otherwise $\varphi(u)$ would also be a left zero).

Hence, by the hypothesis of induction, it is possible to choose p such that the word $y_1 y_2 \cdots y_p$ contains a uniform kth-power modulo φ when considered as a word over the alphabet Y. Since the length of the words y_i is constant, this is also a uniform kth-power modulo φ when it is considered as a word of A^{+}. This completes the proof. ∎

To show that van der Waerden's theorem is a particular case of theorem 4.2.2, consider the morphism from A^{+} to its quotient obtained from the congruence generated by the relations $ab \sim a$ for $a, b \in A$. A uniform kth-power modulo φ is just a word w that contains an arithmetic cadence of order k (see Chapter 3); that is,

$$w = w_0 a w_1 a w_2 \cdots a w_k$$

where $w_0 \in A^{*}$, $a \in A$, and the words $w_1, w_2, \ldots, w_{k-1}$ are all in A^{*} and of the same length.

4.3. Repetitive Semigroups

J. Justin introduced the concept of a *repetitive semigroup* and developed the related theory. (A semigroup S is said to be *repetitive* if for any finite alphabet A any morphism from A^{+} to S is repetitive.)

Theorems 4.2.1 and 4.2.2 are proved by Justin (1972a) in a slightly different form (see Problems 4.3.1 and 4.3.2).

The theory developed by Justin goes much further than the results here (see Justin 1969, 1970, 1971a, 1971b, 1971c, 1972a, 1972b, 1981).

It is impossible to present in this space a complete survey of the theory. Comments here are limited to one other theorem of Justin's; in the Problems section some other results are stated.

THEOREM 4.3.1 (Justin 1972b). *A commutative semigroup is repetitive iff it contains no subsemigroup isomorphic to the commutative free semigroup on two generators.*

The fact that the direct product $\mathbb{P} \times \mathbb{P}$ ($\mathbb{P}$ is the semigroup of positive integers) is not repetitive (the "only if" part) is a consequence of the existence of arbitrarily long words on an alphabet containing two letters without abelian fifth power. (See Justin 1972b and also Chapter 2 for discussion of relations with the problem of words without squares.)

The proof of the "if" part uses another result of Justin's. It is a deep theorem with a very long and technical proof about a "semigroup having bounded generations" (see Justin 1969, 1970); there is also a slightly simpler, but again very long and technical, proof in Pirillo (1981). The complete proof of Theorem 4.3.1 therefore cannot be present here. It is quite general, containing Theorem 4.2.1 as a particular case.

Problems

Section 4.1

4.1.1 Let A be a finite alphabet and w an infinite word on A (see Chapter 2). Prove that there exists an element $a \in A$ and a positive integer p such that for every positive integer k there are positive integers $i_1, i_2, \ldots, i_k$ with

$$w(i_j) = a$$

for every $j, 1 \leqslant j \leqslant k$, and

$$i_{j+1} - i_j \leqslant p$$

for every $j, 1 \leqslant j \leqslant k - 1$. (See Brown 1969.)

Section 4.2

4.2.1. Let A be a finite alphabet, S a finite semigroup, $\varphi: A^+ \to S$ a morphism, and w an infinite word on A. Prove that there exists an idempotent $e \in S$ and a positive integer p such that for every positive integer k the infinite word w contains a kth-power modulo φ whose components $w_1, w_2, \ldots, w_k$ are such that

$$\varphi(w_j) = e \qquad \text{and} \qquad |w_j| \leqslant p$$

for every j, $1 \leqslant j \leqslant k$ (See Brown 1971 and Justin 1972a.)

4.2.2. Given a morphism $\varphi: A^+ \to S$ from A^+ to a finite semigroup S, each sufficiently long word contains a uniform kth power modulo φ whose components have an idempotent of S as image under φ (use Theorem 4.2.2).

Section 4.3

4.3.1. Prove that the (semi)group $\mathbb{Z}$ of integers is repetitive. (See Justin 1972a.)

4.3.2. Prove that the direct product $\mathbb{Z} \times S$, where $\mathbb{Z}$ is the (semi)group of integers and S is a finite semigroup, is repetitive. (See Justin 1972a.)

*4.3.3. Prove that the bicyclic semigroup (i.e. the quotient of $\{a, b\}^*$ by the congruence generated by $ab \sim 1$) is repetitive. (This is a particular case of a more general result in Justin 1971b.)

CHAPTER 5

Factorizations of Free Monoids

5.0. Introduction

The aim of this chapter is to study decompositions of words as a unique ordered product of words taken out of some special sets. More precisely, a factorization of the free monoid A^* is a family $(X_i)_{i \in I}$ of subsets of A^+ indexed by a totally ordered set I such that each word $w \in A^+$ may be written in a unique way as

$$w = x_1 x_2 \cdots x_n$$

with $x_i \in X_{j_i}$ and

$$j_1 \geqslant j_2 \geqslant \cdots \geqslant j_n.$$

The investigation of factorizations of free monoids can be understood as that of some basis for free monoids. These factorizations are in fact, as we shall see, closely related to some basis in the classical sense of linear algebra: the basis of the free Lie algebras.

The main feature of the study of factorizations is the fact that their definition, which makes use of a multiplicative property of the family $(X_i)_{i \in I}$ has strong connections with additive properties of the family $(X_i)_{i \in I}$. This is Schützenberger's theorem of factorizations (theorem 5.4.1), which is a counterpart for free monoids of the well-known Poincare–Birkhoff–Witt theorem.

Let us begin this chapter by studying a particular factorization that is of fundamental importance: the factorization in Lyndon words. One uses a lexicographic order on the free monoid to define Lyndon words as those primitive words that are minimal in their class of conjugates. This is the additive property of the set of Lyndon words: their union is a set of representatives of the conjugate classes of primitive words. In section 5.1 a number of equivalent definitions of Lyndon words are given, and they are proven to form a factorization of the free monoid (Theorem 5.1.5).

In Section 5.2 another kind of factorizations, called *bisections*, are studied, which correspond to the case where the set I of indices of the family $(X_i)_{i \in I}$

has just two elements. The additive property of these factorizations is the following (Proposition 5.2.4 and Corollary 5.2.5): for any partition (P, Q) of A^+, there exists a unique bisection (X, Y) such that $X \subset P, Y \subset Q$.

Section 5.3 contains an exposition of the theory of free Lie algebras. It is self-contained and does not require any previous knowledge of Lie algebras in general. Exhibited first is a basis for the free Lie algebra over the set A, defined as a subalgebra of the free associative algebra $K\langle A \rangle$; this basis is constructed using Lyndon words. Proved next is the Poincare–Birkhoff–Witt theorem (Theorem 5.3.7), which is a fundamental theorem holding for all free algebras over a field and not only for free Lie algebras. It is used to prove that to any bisection of a free monoid is associated a decomposition of the corresponding free Lie algebra into a direct sum of two submodules (Proposition 5.3.11). Finally the Baker–Campbell–Hausdorff theorem about the logarithm and exponential functions in the algebra of power series is proven; it is used in the following section.

In the final section the theorem of factorizations (Theorem 5.4.1) is proven. It shows how the multiplicative property of a factorization $(X_i)_{i \in I}$ can be transformed into an additive property by making use of the logarithm function. This theorem is illustrated by considering the previously studied factorizations and some new ones, such as Viennot factorizations.

5.1   Lyndon Words

Recall that a lexicographic order on the free semigroup A^+ is given by a total order on the alphabet A extended to words in the following way: For any $u, v \in A^+$, $u < v$ iff either $v \in uA^+$ or

$$u = ras, \qquad v = rbt, \quad \text{with} \quad a < b; a, b \in A; r, s, t \in A^*.$$

This defines a total order on A^+. For future reference, we record two properties of the lexicographic order:

($\mathcal{L}$1) $\forall w \in A^*, u < v \Leftrightarrow wu < wv.$

($\mathcal{L}$2)   if $v \notin uA^*, \forall w, z \in A^*, u < v \Rightarrow uw < vz.$

In this section a fixed lexicographic order is defined on A^+.

By definition a *Lyndon word* is a primitive word that is minimal in its conjugate class. The set of Lyndon words will be denoted as L.

Equivalently $l \in L$ iff:

$$\forall u, v \in A^+, l = uv \Rightarrow l < vu.$$

Example 5.1.1. For $A = \{a, b\}$ and $a < b$, the list of the first Lyndon words is

$$L = \{a, b, ab, aab, abb, aaab, aabb, abbb, aaaab, aaabb, aabab, \dots\}$$

The number of Lyndon words of length n is obviously:

$$\text{Card}(L \cap A^n) = \frac{1}{n} \sum_{a|n} \mu(d)[\text{Card}(A)]^{n/d}$$

since this is the number of conjugate classes of primitive words of length n (cf. Chapter 1, Eq. (1.3.7)).

We shall make use of the following characterization:

PROPOSITION 5.1.2. *A word* $w \in A^+$ *is a Lyndon word iff it is strictly smaller than any of its proper right factors*:

$$w \in L \Leftrightarrow \{\forall v \in A^+, w \in A^+v \Rightarrow w < v\}.$$

Proof. Let w be a Lyndon word and v a proper right factor of w:

$$w = uv, \quad \text{with} \quad u,v \in A^+.$$

It must be first shown that v cannot be a left factor of w. If that were the case, we would have for some $t \in A^+$:

$$w = vt = uv.$$

From this equality it can be deduced (see Chapter 1) that for some $x, y \in A^*$ and $i \geq 0$:

$$u = xy, \qquad t = yx, \qquad v = (xy)^i x;$$

whence

$$w = (xy)^{i+1} x.$$

Since w is primitive, $x \neq 1$; and as w is a Lyndon word, it is smaller than its conjugate $x(xy)^{i+1}$:

$$(xy)^{i+1} \ x < x(xy)^{i+1};$$

Canceling the first x gives by $\mathcal{L}1$:

$$(yx)^{i+1} < (xy)^{i+1},$$

and multiplying on the right by x gives, by $\mathcal{L}2$:

$$(yx)^{i+1} x < (xy)^{i+1} x = w,$$

contradicting the definition of Lyndon words; consequently v cannot be a left factor of $w = uv \in L$. Now the hypothesis $v < uv$ yields by $\mathcal{L}2$ $vu < uv$, a contradiction; we thus have proved that a Lyndon word is smaller than any of its proper right factors.

Conversely, if $w \in A^+$ has this property, for any $u, v \in A^+$ such that $w = uv$, we have the inequalities:

$$w < v < vu,$$

showing that $w \in L$.   ∎

The following result gives a second characterization of Lyndon words:

PROPOSITION 5.1.3. *A word $w \in A^+$ is a Lyndon word iff $w \in A$ or $w = lm$ with $l, m \in L, l < m$. More precisely, if m is the proper right factor of maximal length of $w = lm \in L$ that belongs to L, then $l \in L$ and $l < lm < m$.*

Proof. We first prove the "if" part of the statement. Suppose that $l, m \in L$ with $l < m$. We note first that $lm < m$; in fact either l is a left factor of $m = lm'$ and $m < m'$ implies $lm < lm' = m$, or it follows from £2. Now, if v is a right factor of m, we have $lm < m < v$ by Proposition 5.1.2 since $m \in L$; and if v' is a right factor of l, we have $l < v'$, whence by £2, $lm < v'm$. Hence lm is smaller than any of its proper right factors and $lm \in L$ by 5.1.2.

Conversely, for any $w \in L - A$, let m be its longest proper right factor in L (it exists since the last letter of w is in L). We set $w = lm$; if $l \in A$, the property holds; if not, let v be a proper right factor of l. Since $vm \notin L$, let t be a proper right factor of vm such that $t < vm$; then, if we had $v < t$, we would deduce from $v < t < vm$ the existence of an $s < m$ such that $t = vs$ and s would be a proper right factor of m smaller than m, a contradiction with $m \in L$. Hence $t \leqslant v$ and

$$l < lm < t \leqslant v,$$

proving that $l < v$ and, by Proposition 5.1.1, $l \in L$.

Finally, $l < m$ since $w = lm \in L$ implies $lm < m$, and this concludes the proof.   ∎

Proposition 5.1.3 gives a recursive algorithm to construct Lyndon words; it should be noticed that the same word may be obtained several times, or equivalently that the decomposition of a Lyndon word as a product lm, with $l, m \in L$ and $l < m$ need not be unique. For instance,

$$aabb = (aab)(b) = (a)(abb).$$

The pair (l, m), $l, m \in L$ such that $w = lm$ and m of maximal length will be called the *standard factorization* of $w \in L - A$, denoted as $\sigma(w)$. The proof of the following is left as an exercise (Problem 5.1.5).

PROPOSITION 5.1.4. *Let* $w \in L - A$ *and* $\sigma(w) = (l, m)$ *be its standard factorization. Then for any* $n \in L$ *such that* $w < n$, *the pair* (w, n) *is the standard factorization of* $wn \in L$ *iff* $n \leqslant m$.

The main result of this section is the following:

THEOREM 5.1.5. (Lyndon). *Any word* $w \in A^+$ *may be written uniquely as a nonincreasing product of Lyndon words*:

$$w = l_1 l_2 \cdots l_n, \quad l_i \in L, \quad l_1 \geqslant l_2 \geqslant \cdots \geqslant l_n.$$

Proof. Any word $w \in A^+$ may be written in at least one way as a product of Lyndon words (since the letters are elements of L):

$$w = l_1 l_2 \cdots l_n;$$

choose a factorization with n minimal; then if, for some $i, 1 \leqslant i \leqslant n - 1$, we have $l_i < l_{i+1}$, by Proposition 5.1.3, the product $l_i l_{i+1} \in L$ and therefore n is not minimal. This proves the existence of at least one nonincreasing factorization.

To prove the uniqueness of the factorization, suppose

$$l_1 l_2 \cdots l_n = l_1' l_2' \cdots l_{n'}', \quad l_i, l_i' \in L,$$
$$l_1 \geqslant l_2 \geqslant \cdots \geqslant l_n, \quad l_1' \geqslant l_2' \geqslant \cdots \geqslant l_n'.$$

If, for instance l_1 is longer than l_1', we have:

$$l_1 = l_1' l_2' \cdots l_i' u$$

for some left factor u of l_{i+1}'. Now, by Proposition 5.1.2,

$$l_1 < u \leqslant l_{i+1}' \leqslant l_1' < l_1,$$

a contradiction. Hence, $l_1 = l_1'$, and the result follows by induction on n. ∎

The proof just given supplies an algorithm to factorize a word in Lyndon words. The following proposition gives a faster algorithm:

PROPOSITION 5.1.6. *Let* $w = l_1 l_2 \cdots l_n$ *be the factorization of* $w \in A^+$ *as a nonincreasing product of Lyndon words. Then* l_n *is the smallest right factor of* w.

Proof. Let v be the smallest right factor of w and set $w = uv$; then $v \in L$ by 5.1.2. If $u = 1$, the result is proved. If not, let s be the smallest right factor of u with $w = rsv$. If $s < v$, then by 5.1.2, $sv \in L$, hence $sv < v$, a contradiction with the definition of v. Therefore, $s \geqslant v$ and iterating this process gives the factorization of w.   ∎

Notice that we could also have used the proof of 5.1.6 to prove 5.1.5; also, the algorithm suggested by 5.1.6 is not the best possible. Instead of operating from right to left, it is possible to operate from left to right to obtain the factorization of a word w using a number of comparisons of letters, which is a linear function of the length of w (Duval 1980).

5.2   Bisections of Free Monoids

Let A be a (finite or infinite) alphabet; a pair (X, Y) of subsets of A^+ is a bisection of A^* if any word $w \in A^*$ may be written uniquely as:

$$w = x_1 x_2 \cdots x_r y_1 y_2 \cdots y_s, \qquad (5.2.1)$$

with $x_i \in X$, $y_j \in Y$, and $r, s \geqslant 0$.

Equivalently, (X, Y) is a bisection of A^* iff X and Y are codes and if any word $w \in A^*$ may be written uniquely as:

$$w = xy, \quad x \in X^*, y \in Y^*. \qquad (5.2.2)$$

Example 5.2.1. For $A = \{a, b\}$, the sets $X = a^*b$ and $Y = a$ form a bisection of A^*. Indeed, any word $w \in A^*$, either is in a^*, or may be written uniquely:

$$w = a^{i_1} b a^{i_2} b \cdots a^{i_r} b a^{i_{r+1}}, \quad i_j \geqslant 0, \ r \geqslant 1.$$

Example 5.2.2. Consider, for $A = \{a, b\}$, the congruence of A^* generated by the relation:

$$ab \sim 1.$$

The class of 1 is a submonoid (called the Dyck language—cf. Chapter 11) which is free; denote as D its basis and set:

$$X = D^*b, \quad Y = D \cup a.$$

The pair (X, Y) will be proved to be a bisection of A^*. For instance

$$w = [abaababbabb][aba]$$

may be factorized as indicated by the brackets.

The definition of a bisection may be reformulated using characteristic series. A pair (X, Y) of nonempty subsets of A^* is obviously a bisection iff:

$$\mathbf{A}^* = (\mathbf{X})^*(\mathbf{Y})^*. \tag{5.2.3}$$

By taking the inverses, we obtain the equality

$$\mathbf{YX} + \mathbf{A} = \mathbf{X} + \mathbf{Y}. \tag{5.2.4}$$

The equality (5.2.4) means that (X, Y) is a bisection iff X, Y are codes and:

(i) $YX \cup A \subset X \cup Y$
(ii) $X \cap Y = \varnothing$
(iii) any $w \in (X \cup Y) - A$ may be written uniquely as

$$w = yx, \qquad y \in Y, \qquad x \in X.$$

This suggests a method for constructing bisections that consists in distributing the elements of A between X and Y and then recursively distributing the elements of $YX \cap A^n$, for each $n \geqslant 2$, between X and Y. This method will be developed later (see Corollary 5.2.5).

Example 5.2.3. The equality of type (5.2.4) corresponding to the bisection of Example 5.2.1 is

$$aa^*b + a + b = a^*b + b,$$

which obviously holds.

To prove that $(D^*b, D \cup a)$ is a bisection (Example 5.2.2), we compute

$$(\mathbf{D} + a)\mathbf{D}^*b + a + b = \mathbf{DD}^*b + a\mathbf{D}^*b + a + b$$
$$= \mathbf{D}^*b + a\mathbf{D}^*b + a$$

since $\mathbf{DD}^* + 1 = \mathbf{D}^*$. Now, we have $\mathbf{D} = a\mathbf{D}^*b$, as may be verified (Problem 5.2.3 or Chapter 11) and therefore,

$$(\mathbf{D} + a)\mathbf{D}^*b + a + b = \mathbf{D}^*b + \mathbf{D} + a,$$

proving the Eq. (5.2.4) holds.

The following proposition shows that, in order that (X, Y) be a bisection, it is enough that is satisfies a condition seemingly weaker than Eq. (5.2.4):

PROPOSITION 5.2.4. *Let X, Y be two disjoint subsets of A^+. Then (X, Y) is a bisection of A^* iff*

$$YX \cup A = X \cup Y. \tag{5.2.5}$$

Proof. The condition is necessary since it is a consequence of Eq. (5.2.4) Conversely if (5.2.5) holds we proceed by steps; first we have

$$A^* = X^*Y^*. \tag{5.2.6}$$

In fact each word $w \in A^+$ has at least one factorization

$$w = z_1 z_2 \cdots z_n$$

with $z_i \in X \cup Y$ since $A \subset X \cup Y$. If n is chosen minimal, then $z_i \in Y, 1 \leqslant i \leqslant n - 1$, implies $z_{i+1} \in Y$ since $YX \subset X \cup Y$. This proves (5.2.6). We now proceed to prove that the factorization is unique.

Let us first show that for any $u, v \in A^*$,

$$uv \in X \Rightarrow v \in X^*. \tag{5.2.7}$$

We prove (5.2.7) by induction on $n = |uv|$. If $n = 1$ it is obvious. Furthermore, if (5.2.7) holds up to $n - 1$, let $x = uv \in X \cap A^n$; by (5.2.5), we may write $x = y_1 x_1$, $y_1 \in Y$, $x_1 \in X$ and rewriting again y_1 gives finally

$$x = y_k x_k x_{k-1} \cdots x_1 \quad \text{with} \quad y_k \in Y \cap A, x_i \in X.$$

We then have $v = rx_{p-1} \cdots x_1$ with $1 \leqslant p \leqslant k$ and r a right factor of x_p. By induction hypothesis, we have $r \in X^*$ and this proves (5.2.7). Symmetrically, we can prove:

$$uv \in Y \Rightarrow u \in Y^*. \tag{5.2.8}$$

We can now show that

$$X^* \cap Y^* = 1. \tag{5.2.9}$$

In fact, in view of (5.2.7), (5.2.8), it is enough to prove that

$$X \cap Y^* = \varnothing, \qquad X^* \cap Y = \varnothing. \tag{5.2.10}$$

Suppose that no word of A^* of length at most $n - 1$ belongs to $X \cap Y^*$ or $X^* \cap Y$; if $w \in A^n \cap X^* \cap Y$, write $w = x_1 x_2 \cdots x_p$, $x_i \in X$. Then $p \geqslant 2$ since $X \cap Y = \varnothing$. But, by (5.2.8), $x_1 \in Y^*$, a contradiction.

Let us now show that X and Y are codes. In fact, in view of (5.2.7), it is enough to prove that $X \cap X^r = \varnothing$ for $r \geqslant 2$. We proceed by induction on $|x|$: suppose that $x = x_1 x_2 \cdots x_r \in X$ with $x_i \in X$; then write $x = y'x'$ with $y' \in Y, x' \in X$. By (5.2.8) and (5.2.10), x_1 cannot be a left factor of y'; therefore $x_1 = y'u$ with $u \in A^+$. By (5.2.7), $x' = ux_2 \cdots x_r$ is in $X^2 X^*$, a contradiction with the induction hypothesis.

Finally, we obtain the uniqueness of the factorization of a word $w \in A^+$ as

$$w = x_1 \cdots x_r y_1 \cdots y_s \quad x_i \in X, \ y_j \in Y.$$

Since $xy = x'y', x, x' \in X^*$ implies $x = x', y = y'$ by (5.2.7)–(5.2.9). ∎

As an immediate consequence of Proposition 5.2.4 we have the following:

COROLLARY 5.2.5. *Let (P, Q) be a partition of A^+. There exists a unique bisection (X, Y) of A^* such that:*

$$X \subset P, \qquad Y \subset Q.$$

This corollary shows that in order to construct a bisection (X, Y) of A^* it is enough to share the elements of A between X and Y and, recursively, to share the elements of $YX \cap A^n$, for each $n \geq 2$, between X and Y.

Example 5.2.5. For $A = \{a, b\}$, construct X and Y as the preceding corollary by choosing to assign all elements of YX, whose length is at least 4, to X and to distribute the first ones as in the following array:

n	X	Y
1	a	b
2	ba	
3		b^2a
≥ 4	R	

One has

$$\mathbf{R} = b^2a(a + ba) + (b + b^2a)\mathbf{R},$$

whence

$$R = \{b, b^2a\}^* \{b^2a^2, b^2aba\}.$$

The corresponding bisection is

$$X = \{a, ba\} \cup \{b, b^2a\}^* \{b^2a^2, b^2aba\}$$
$$Y = \{b, b^2a\}.$$

The factorization of a word w may be obtained by considering the graph of Figure 5.1. For any word $w \in A^*$, there is a unique path with label w starting at vertex 1, and the largest left factor of w leading back to vertex 1 gives the left factor of w in X^* of its decomposition (this is a small incursion in automata theory (see Eilenberg 1974 for a method allowing derivation of graph shown in Figure 5.1 from the expression of X and Y).

The following result provides an alternative construction of bisections that uses the graph shown in Example 5.2.5.

PROPOSITION 5.2.7. *A pair* (X, Y) *of subsets of* A^+ *is a bisection iff there exists a partially ordered set* Q *with a maximal element* q^+ *and a morphism* α *of* A^* *into the monoid of order-preserving mappings of* Q *into* Q; *that is,*

$$q \leqslant q' \Rightarrow q\alpha(w) \leqslant q'\alpha(w),$$

such that X *is the basis of the submonoid*

$$X^* = \{x \in A^* \mid q^+\alpha(x) = q^+\}$$

and Y *the basis of* $Y^* = A^* - XA^*$.

Proof. If (X, Y) is a bisection of A^*, let Q be the family of subsets of A^*

$$u^{-1}X^* = \{v \in A^* \mid uv \in X^*\}$$

for all $u \in A^*$. We order these subsets by inclusion. Then

$$q^+ = X^*$$

is a maximal element since by (5.2.7):

$$uv \in X^* \Rightarrow v \in X^*.$$

Now, for each $w \in A^*$, we define a mapping $\alpha(w)$ of Q into Q by

$$(u^{-1}X^*)\alpha(w) = (uw)^{-1}X^*.$$

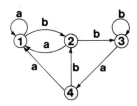

Figure 5.1. Unique path for factorization of a word w.

First α is a well-defined mapping of A^* into Q^Q since

$$u^{-1}X^* = v^{-1}X^* \Rightarrow (uw)^{-1}X^* = (vw)^{-1}X^*,$$

and it is a monoid morphism.

For each $w \in A^*$, this mapping is order-preserving since:

$$u^{-1}X^* \subset r^{-1}X^* \Rightarrow (uw)^{-1}X^* \subset (rw)^{-1}X^*.$$

Finally $x \in X^*$ iff $q^+\alpha(x) = q^+$; and $Y^* = A^* - XA^*$ as in any bisection.

Conversely, if Q satisfies the hypothesis of Proposition 5.2.6, any word $w \in A^+$ may be factorized in a unique way as

$$w = xy \qquad\qquad (5.2.11)$$

where $q^+\alpha(x) = q^+$ and for any left factor $u \neq 1$ of y, $q^+ . u \neq q^+$. Now the set

$$\{x \in A^* \mid q^+\alpha(x) = q^+\}$$

is a free submonoid and let X be its basis. The set

$$P = \{y \in A^* \mid y = uv \quad u \neq 1, \Rightarrow q^+\alpha(u) \neq q^+\}$$

is also a submonoid; in fact if $y, z \in P$, then

$$q^+\alpha(y) < q^+ \Rightarrow q^+\alpha(y, z) \leqslant q^+\alpha(z) < q^+$$

so that $yz \in P$. This submonoid is free since

$$uv \in P \Rightarrow u \in P;$$

if Y is its basis, then (X, Y) is a bisection by (2.11).   ∎

Example 5.26. Continued. The set Q may be identified with the set of vertices of the graph of Figure 5.1 with the order $3 < 2 < 1$ and $4 < 1$; the mapping α is defined by $q\alpha(w) = q'$ iff the path with label w beginning at q ends at q'.

To end this section here is an interesting example of bisections.

Example 5.2.8. Consider a morphism σ from A^* into the additive group $\mathbb{Z}$ of rational integers. Define two sets P and N as:

$$P = \{p \in A^* \mid \sigma(p) > 0\} \cup 1, \qquad N = \{n \in A^* \mid \sigma(n) \leqslant 0\},$$

and further define

$$R = \{r \in A^* \mid r = uv \Rightarrow v \in P\}, \qquad S = \{s \in A^* \mid s = uv \Rightarrow u \in N\}.$$

The set R (resp. S) is a submonoid containing all right (resp. left) factors of its elements; let X (resp. Y) be its basis.

The pair (X, Y) is a bisection of A^* since for any word $w \in A^*$, we have

$$w = rs, \qquad r \in R, \qquad s \in S,$$

where r is the smallest left factor of w such that $\sigma(r)$ is maximal among the $\sigma(u)$, for all left factors u of w.

In the case of the morphism defined by $\sigma(a) = -1, \sigma(b) = +1$, this bisection is the same as that of Example 5.2.2 (see Problem 5.2.3). This factorization may be graphically obtained as on Figure 5.2.

This example presents an interesting property, which gives a combinatorial interpretation of the so-called Sparre-Andersen equivalence principle (Foata and Schützenberger 1971).

Let us recall this principle in a simple way. Associate to each sequence $s = (x_1, \ldots, x_n) \in \mathbb{Z}^n$ of n rational integers two quantities:

$L(s) =$ the number of strictly positive partial sums $s_k = x_1 + x_2 + \cdots + x_i$.

$\Pi(s) =$ the index of the first maximum among these partial sums.

Sparre-Andersen has discovered the surprising fact that, among the permutations of a given sequence s, the two numbers L and Π have the same distribution. In terms of probabilities, this means that the two random variables (defined on sequences of fixed length of respectively independent

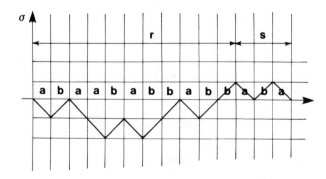

Figure 5.2. $\sigma(a) = -1, \sigma(b) = +1$.

real random variables) equal respectively to the index of the first maximum and the time spent above the x-axis, have the same distribution.

To see how this is related to the preceding factorization, associate to each word $w \in A^+$ two integers

$$L(w) = \mathrm{Card}\{u \in P - 1 \mid w \in uA^*\},$$

which is the number of nonempty left factors of w in P, and

$$\Pi(w) = |x|$$

if $w = xy$, $x \in X^*$, $y \in Y^*$. These numbers are obviously the corresponding ones for the sequence

$$s = (\sigma(a_1), \ldots, \sigma(a_n))$$

where $w = a_1 a_2 \cdots a_n$, $a_i \in A$.

To give a proof of the equivalence principle it is therefore sufficient to exhibit a one-to-one mapping of A^+ into itself preserving the value of σ and exchanging that of L and Π. For this purpose, define a transformation ρ on A^* fixing the letters $a \in A$ and inductively computed as:

$$\rho(wa) = a\rho(w) \quad \text{if} \quad aw \in P,$$
$$\rho(wa) = \rho(w)a \quad \text{if} \quad wa \in N.$$

It is one-to-one since its inverse is the transformation defined by

$$\tau(aw) = \tau(w)a \quad \text{if} \quad aw \in P,$$
$$\tau(aw) = a\tau(w) \quad \text{if} \quad aw \in N.$$

PROPOSITION 5.2.9. *For every word $w \in A^+$, one has $L(w) = \Pi(\rho w)$.*

Proof. Let u be the left factor of maximal length of w that belongs to N or P depending upon $w \in P$ or $w \in N$. If $w = uv$, we have as a consequence of the definition of ρ that

$$\rho(w) = \tilde{v}\rho(u) \quad \text{if} \quad w \in P,$$
$$\rho(w) = \rho(u)v \quad \text{if} \quad w \in N,$$

where $\tilde{v}$ is the reverse of v. Now, by definition, we have

$$\tilde{v} \in X^* \quad \text{if} \quad w \in P,$$
$$v \in Y^* \quad \text{if} \quad w \in N.$$

This proves the property by induction on $|w|$ since:

$$L(w) = L(u) + |v|, \qquad \Pi\rho(w) = |v| + \Pi(\rho u) \quad \text{if} \quad w \in P,$$
$$L(w) = L(u), \qquad \Pi(w) = \Pi(\rho u) \quad \text{if} \quad w \in N. \qquad \blacksquare$$

This problem will be considered in a more general framework in Chapter 10.

5.3. Free Lie Algebras

Let K be a commutative ring with unit; a *Lie algebra* $\mathfrak{L}$ (over K) is a K-algebra whose product, denoted with brackets, satisfies the two identities:

$$[xx] = 0, \tag{5.3.1}$$
$$[[xy]z] + [[yz]x] + [[zx]y] = 0. \tag{5.3.2}$$

The identity (5.3.1) implies

$$[xy] + [yx] = 0 \tag{5.3.3}$$

and the identity (5.3.2) is called the *Jacobi identity*. It may be usefully written as

$$[[xy]z] = [x[yz]] - [y[xz]], \tag{5.3.4}$$

which is equivalent to (5.3.2) using (5.3.3) and can be viewed as a "rewriting rule" to collect the brackets on the right side.

For any associative K-algebra R, the product

$$[xy] = xy - yx, \quad x, y \in R, \tag{5.3.5}$$

turns the K-module R into a Lie algebra denoted as R_L.

When R is the free associative algebra $K\langle A \rangle$ over the set A, the subalgebra of R_L generated by A is called the *free Lie algebra* over A, denoted as $\mathfrak{L}_K(A)$ or $\mathfrak{L}(A)$ (the adjective *free* will be justified later on).

The module $\mathfrak{L}(A)$ is generated by those of its elements that are homogeneous polynomials of R. We denote as R_n (resp $\mathfrak{L}_n(A)$) the submodule of R (resp. $\mathfrak{L}(A)$) generated by homogeneous elements of degree $n(n \geqslant 0)$. Therefore

$$\mathfrak{L}(A) = \bigoplus_{n \geqslant 1} \mathfrak{L}_n(A), \tag{5.3.6}$$

and

$$\mathfrak{L}_{n+1}(A) = [\mathfrak{L}_n(A), \mathfrak{L}_1(A)], \tag{5.3.7}$$

where, for any two submodules M, N of a Lie algebra $\mathfrak{L}$, the symbol $[M, N]$ denotes the submodule generated by the elements $[m, n]$, $m \in M$, $n \in N$. It is easy to verify that:

$$\mathfrak{L}_0(A) = 0 \tag{5.3.8}$$

and

$$\mathfrak{L}_1(A) = R_1. \tag{5.3.9}$$

It will now be proved that the K-module $\mathfrak{L}(A)$ is free, and a basis will be shown.

Choose an order on the set A and let L be the set of Lyndon words over the alphabet A (cf. Section 5.1). We define a mapping λ of L into $\mathfrak{L}(A)$ inductively by

$$\lambda(a) = a, \quad a \in A, \tag{5.3.10}$$

and

$$\lambda(l) = [\lambda(m), \lambda(n)]$$

if $l \in L - A$ and $\sigma(l) = (m, n)$ is the standard factorization of l (see Proposition 5.1.4).

THEOREM 5.3.1. *The K-module $\mathfrak{L}(A)$ is free with $\lambda(L)$ as a basis.*

The proof relies on two lemmas.

LEMMA 5.3.2. *For each $l \in L \cap A^k$, $k \geqslant 1$, one has $\lambda(l) = l + r$, where r belongs to the submodule of $K\langle A \rangle$ generated by those words $w \in A^k$ such that $l < w$.*

Proof. The property is true for $k = 1$, and it can be proved by induction on k: let $k \geqslant 2$, $l \in L \cap A^k$ and $\sigma(l) = (m, n)$. Then $\lambda(l) = [\lambda(m), \lambda(n)]$. By induction hypothesis $\lambda(m) = m + r$, $\lambda(n) = n + s$, where r and s belong, respectively, to the submodules generated by $M = \{w \in A^i | w > m\}$ and $N = \{w \in A^j | w > n\}$ with $i = |m|$, $j = |n|$. Then

$$\lambda(l) = mn + ms + r(n + s) - nm - nr - s(m + r).$$

It can be verified that each term, except mn, in the right-hand side belongs to the submodule generated by $P = \{w \in A^k | w > l\}$; for the term nm this is a consequence of the definition of L and for the others of lexicographic order. This proves the lemma. ∎

To state the next lemma, we introduce an order on the sets

$$X_k = \{(m, n) \in L \times L \mid m < n, \, mn \in A^k\}, \quad k \geqslant 2.$$

By definition, $(m, n) < (m', n')$ if $mn > m'n'$ (note the reversed order) or if $mn = m'n'$ and $m < m'$.

LEMMA 5.3.3. *For each $(m, n) \in X_k$, the element $[\lambda(m), \lambda(n)]$ belongs to the submodule of $\mathfrak{L}(A)$ generated by the elements $\lambda(l)$ with $l \in L \cap A^K$ and $\sigma(l) \leqslant (m, n)$.*

Proof. The lemma is proved by a double induction, first on k and then on the order defined above on X_k. The minimal elements for this order are the pairs of letters, for which the result obviously holds.

It may be supposed, by the induction hypothesis, that the result holds for all pairs $(u, v) \in X_j$ for $j < k$ or $j = k$ and $(u, v) < (m, n)$.

First, if $m \in A$, the factorization (m, n) is standard and therefore

$$[\lambda(m), \lambda(n)] = \lambda(mn).$$

Further, if $m \in I \quad A$, let $\sigma(m) = (u, v)$. If $n \leqslant v$ then, by Proposition 5.1.4, the factorization (m, n) is again standard. Supposing that $v < n$, therefore, compute:

$$\begin{aligned}
[\lambda(m), \lambda(n)] &= [[\lambda(u), \lambda(v)], \lambda(n)] \\
&= [\lambda(u), [\lambda(v), \lambda(n)]] - [\lambda(v), [\lambda(u), \lambda(n)]].
\end{aligned}$$

$$(5.3.11)$$

The two terms of the right-hand side of (5.3.11) are treated separately. For the first one, by induction hypothesis (since $|vn| < k$):

$$[\lambda(v), \lambda(n)] = \sum_i \alpha_i \lambda(w_i)$$

with $\alpha_i \in K, w_i \in L$ and $\sigma(w_i) \leqslant (v, n)$. Hence,

$$[\lambda(u), [\lambda(v, \lambda(n)]] = \sum_i \alpha_i [\lambda(u), \lambda(w_i)].$$

Now $u < uvn < vn \leqslant w_i$ and therefore $u < w_i$. As $vn \leqslant w_i$, we obtain $uvn \leqslant uw_i$ and $(u, w_i) < (uv, n) = (m, n)$; by induction hypothesis, this implies

$$[\lambda(u), [\lambda(v), \lambda(n)]] = \sum_{i, j} \beta_{ij} \lambda(w_{ij})$$

with $w_{ij} \in L$ and $\sigma(w_{ij}) \leqslant (u, w_i) < (uv, w) = (m, n)$.

For the second term the argument is similar:

$$[\lambda(u), \lambda(n)] = \sum_i \alpha_i \lambda(w_i)$$

with $w_i \in L$ and $\sigma(w_i) \leqslant (u, n)$. For each index i, either $v < w_i$ and then $vw_i \geqslant vun > uvn$, the last inequality resulting from $uv \in L$; either $v = w_i$ and $[\lambda(v), \lambda(w_i)] = 0$; or $w_i < v$ and $w_i v \geqslant unv > uvn$, since $vn \in L$. In all cases, we obtain by induction hypothesis

$$[\lambda(v), \lambda(w_i)] = \sum_j \alpha_{ij} \lambda(w_{ij})$$

with $w_{ij} \in L$ and $w_{ij} \geqslant vw_i > uvn$ for the case $v < w_i$, $w_{ij} \geqslant w_i v > uvn$ for the case $v > w_i$. Therefore the second term of the right-hand side of Eq. (5.3.11) also belongs to the submodule generated by the $\lambda(w)$ with $\sigma(w) < (m, n)$, and this concludes the proof of the lemma. ■

It is easy to deduce Theorem 5.3.1 from the lemmas. In fact, $\lambda(L)$ is a set of linearly independent elements since Lemma 5.3.2 implies that the projection of the submodule generated by $\lambda(L)$ into the submodule generated by L is injective. Further, the submodule generated by $\lambda(L)$ is, by Lemma 5.3.3, a subalgebra of $K\langle A \rangle_L$; since it contains A, it is equal to $\mathfrak{L}(A)$. This proves Theorem 5.3.1.

Example 5.3.4. As an illustration of the algorithm underlying Lemma 5.3.3: Let $A = \{a, b, c\}$ with $a < b < c$, and consider $ab, c \in L$; then $\lambda(ab) = [ab]$ and

$$[[ab]c] = [a[bc]] - [b[ac]]$$
$$= [a[bc]] + [[ac]b]$$

where $abc, acb \in L$ and $\sigma(abc) = (a, bc)$, $\sigma(acb) = (ac, b)$.

COROLLARY 5.3.5 (Witt's formula). *The dimension of $\mathfrak{L}_n(A)$ is*

$$\psi_k(n) = \frac{1}{n} \sum_{d|n} \mu(d) k^{n/d}$$

with $k = \mathrm{Card}\ (A)$.

In fact, the dimension of $\mathfrak{L}_n(A)$ is, by Theorem 5.3.1. equal to the number of Lyndon words of length n.

There is of course a striking analogy between the fact that the set L of Lyndon words is a factorization of the free monoid and the fact that $\lambda(L)$ is a base of $\mathcal{L}(A)$. This is closely related to the classical theorem of Poincaré, Birkhoff, and Witt that is proven subsequently.

Given a Lie algebra $\mathcal{L}$, we define an *enveloping algebra* for $\mathcal{L}$ as an associative algebra U and a morphism φ from $\mathcal{L}$ into U_L such that for each morphism ψ from $\mathcal{L}$ into S_L, where S is any associative algebra, there exists a unique morphism, θ from U into S such that the following diagram is commutative:

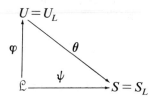

The enveloping algebra of $\mathcal{L}$ is obviously unique (up to isomorphism). It is not difficult to prove directly its existence. However we shall obtain it as a consequence of an explicit construction that provides more information on the enveloping algebra.

Let us make the hypothesis that $\mathcal{L}$ is a free K-module. This is true when $\mathcal{L}$ is the free Lie algebra $\mathcal{L}(A)$ for any ring K, by Theorem 5.3.1, and it is always true when K is a field.

Let E be a basis for the K-module $\mathcal{L}$ and consider a one-to-one mapping φ from E onto a set B

$$\varphi \colon E \to B.$$

The mapping φ extends to an isomorphism from $\mathcal{L}$ onto the free K-module $K[B]$ on the set B. This defines a structure of Lie algebra on $K[B]$ by defining $[bc] = \varphi([ef])$ for $\varphi(e) = b, \varphi(f) = c$.

We order the set B and we consider in the free monoid B^* the set

$$F = \{b_1 b_2 \cdots b_n \mid b_i \in B, \quad n \geq 0, \quad b_1 \geq b_2 \geq \cdots \geq b_n\}$$

of nonincreasing words over the alphabet B.

Define the *index* $\nu(w)$ of a word $w \in B^*$ as the number of triples (r, s, t) of words in B^* such that

$$w = rasbt$$

with $a, b \in B$ and $a < b$. Clearly, for any $u, v \in B^*$ and $a, b \in B$ with $a < b$,

$$\nu(ubav) = \nu(uabv) - 1.$$

Also, $\nu(w) = 0$ iff $w \in F$.

LEMMA 5.3.6. *There exists a unique endomorphism* λ *of the module* $K\langle B\rangle$ *such that*

(i) $\lambda(f) = f$, if $f \in F$
(ii) $\lambda(ubcv) = \lambda(ucbv) + \lambda(u[bc]v)$, $\quad u, v \in B^*$, $\quad b, c \in B$.

Moreover, λ *maps* $K\langle B\rangle$ *onto* $K[F]$ *and*

(iii) $\lambda(uv) = \lambda(\lambda(u)v) = \lambda(u\lambda(v))$, $\quad$ for $\quad u, v \in K\langle B\rangle$.

Proof. The uniqueness of λ is obvious since, by condition (i), λ is defined on F (which contains B) and if $w = ubcv$ with $u, v \in B^*$, $b < c$, in the right-hand side of (ii) λ is applied to a word of smaller index for the first term and of smaller length for the second. This also shows that λ maps $K\langle B\rangle$ onto $K[F]$.

For the existence of λ, use an induction first on the length and second on the index. Suppose that a mapping λ satisfying (i) and (ii) has already been defined for the words of length at most $n - 1$ and the words of length n with index at most $v - 1$. Let $w \in B^*$ be such that $|w| = n$, $v(w) = v$. We show that for each pair of factorizations $w = ubcv = u'b'c'v'$, with $u, u', v, v' \in B^*$, $b, c, b', c' \in B$, $b < c$, $b' < c'$,

$$\lambda(ucbv) + \lambda(u[bc]v) = \lambda(u'c'b'v') + \lambda(u'[b'c']v'). \qquad (5.3.12)$$

We may assume that $|u| < |u'|$. If $|u| \leq |u'| - 2$, then we have $u' = ubct$ with $t \in B^*$. By the induction hypothesis, the common value of the two sides of (5.3.12) is then

$$\lambda(ucbtc'b'v') + \lambda(ucbt[b'c']v') + \lambda(u[bc]tc'b'v') + \lambda(u[bc]t[b'c']v').$$

Next, if $|u| = |u'| - 1$, we have $u' = ub$, $c = b'$, $v = c'v'$. We therefore have to prove that whenever $w = rabcs$ with $r, s \in B^*$, $a, b, c \in B$, $a < b < c$, then

$$\lambda(rbacs) + \lambda(r[ab]cs) = \lambda(racbs) + \lambda(ra[bc]s). \qquad (5.3.13)$$

By the induction hypothesis, the left-hand side of (5.3.13) may be rewritten as

$$\lambda(rcbas) + \lambda(r[bc]as) + \lambda(rb[ac]s) + \lambda(r[ab]cs) \qquad (5.3.14)$$

and the right-hand side of (5.3.13) may be rewritten as

$$\lambda(rcbas) + \lambda(rc[ab]s) + \lambda(r[ac]bs) + \lambda(ra[bc]s). \qquad (5.3.15)$$

The difference between (5.3.14) and (5.3.15) is equal to

$$\lambda(r[bc]as) - \lambda(ra[bc]s) + \lambda(rb[ac]s) - \lambda(r[ac]bs)$$
$$+ \lambda(r[ab]cs) - \lambda(rc[ab]s).$$

By induction hypothesis, this can be rewritten as

$$\lambda(r[[bc]a]s) + \lambda(r[b[ac]]s) + \lambda(r[[ab]c]s).$$

But by the Jacobi identity,

$$[[bc]a] + [b[ac]] + [[ab]c] = 0$$

and therefore (5.3.13) holds.

Now $\lambda(w)$ can be defined as the common value of the two sides of (5.3.12), for all the words of length n and index v. This mapping satisfies conditions (i) and (ii), and this proves the existence of λ. Finally, an easy induction shows that λ satisfies condition (iii). ∎

We now define an algebra U by defining on the module $K[F]$ the following product:

$$u * v = \lambda(uv),$$

where the product uv is that of $K\langle B \rangle$. The algebra U is associative since, using condition (iii) of Lemma 5.3.6

$$u * (v * w) = \lambda(u\lambda(vw)) = \lambda(uvw) = \lambda(\lambda(uv)w) = (u * v) * w.$$

Since U contains $K[B]$ there is a morphism of modules

$$\varphi: \mathcal{L} \to U.$$

This is a morphism of algebras from $\mathcal{L}$ into U_L since, for $e, f \in E$, there are, with $\varphi(e) = b, \varphi(f) = c$, the equalities

$$\varphi([ef]) = [bc] = b * c - c * b.$$

THEOREM 5.3.7. (Poincaré, Birkhoff, Witt). (U, φ) is the enveloping algebra of $\mathcal{L}$.

Proof. Let ψ be a morphism from $\mathcal{L}$ into S_L, where S is an associative algebra. Then any morphism θ from U into S such that $\psi = \theta\varphi$ must satisfy:

$$\theta(b) = \psi(e) \qquad\qquad (5.3.16)$$

for each $b \in B$, with $\varphi(e) = b$. This proves the uniqueness of θ since B generates U.

To prove the existence of θ, let us define θ on B by (5.3.16) and then multiplicatively on F. Then, for each $r, s \in F$,

$$\theta(r*s) = \theta(r)\theta(s).$$

In fact, either $rs \in F$ and this follows from the definition of θ, or $r = ub, s = cv$ with $u, v \in B^*$, $b, c \in B$, $b < c$. Then

$$r*s = \lambda(rs) = \lambda(ucbv) + \lambda(u[bc]v).$$

Hence

$$\theta(r*s) = \theta(\lambda(ucbv)) + \theta(\lambda(u[bc]v)).$$

Now we may suppose, by induction first on $|w|$, second on the index $\nu(w)$ that:

$$\theta\lambda(ucbv) = \theta(u)\theta(cb)\theta(v),$$
$$\theta\lambda(u[bc]v) = \theta(u)\theta[bc]\theta(v).$$

Since ψ is a morphism from $\mathcal{L}$ into S_L,

$$\theta(r*s) = \theta(u)\theta(b)\theta(c)\theta(v)$$
$$= \theta(r)\theta(s)$$

and this concludes the proof. ∎

The first consequence of Theorem 5.3.7 is a fact that was not obvious at all from the definition of the enveloping algebra.

COROLLARY 5.3.8. *The canonical morphism φ from a Lie algebra $\mathcal{L}$ into its enveloping algebra U is injective.*

Proof. This is obvious since φ is an isomorphism of $\mathcal{L}$ into $U = K[F]$. ∎

The second corollary justifies the name of free Lie algebra given to the subalgebra $\mathcal{L}(A)$ of $K\langle A \rangle_L$ generated by A.

COROLLARY 5.3.9. *The free algebra $K\langle A \rangle$ is the enveloping algebra of $\mathcal{L}(A)$ and for any mapping β of the set A into a Lie algebra $\mathcal{L}$, there exists a unique morphism γ from $\mathcal{L}(A)$ into $\mathcal{L}$ extending β.*

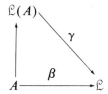

Proof. Let R be the enveloping algebra of $\mathfrak{L}(A)$, that we can regard by Corollary 3.1.7 as a submodule of R. As $\mathfrak{L}(A)$ is a subalgebra of $K\langle A\rangle_L$, there exists a unique morphism of R into $K\langle A\rangle$ that is the identity on A. This implies $R = K\langle A\rangle$.

Next, if β is a mapping from A into a Lie algebra $\mathfrak{L}$, let S be the enveloping algebra of $\mathfrak{L}$. Then, by 5.3.7, $\mathfrak{L}$ can be considered as a submodule of S and therefore β extends uniquely to a morphism γ from $K\langle A\rangle$ into S, which maps $\mathfrak{L}(A)$ into $\mathfrak{L}$.   ∎

Remark 5.3.10. It is possible to derive Witt's formula (Corollary 5.3.5) from Theorem 5.3.7 admitting that $\mathfrak{L}(A)$ is a free K-module. In fact let E be a basis of $\mathfrak{L}(A)$ which is supposed to be totally ordered and formed of homogeneous elements. We denote by $|e|$ the degree of $e \in E$.

Then the elements

$$e_1 e_2 \cdots e_n, \qquad e_1 \geqslant e_2 \geqslant \cdots \geqslant e_n$$

form a basis of $K\langle A\rangle$. The dimension k^n (with $k = \operatorname{Card}(A)$) of the module of homogeneous polynomials of degree n is therefore also equal to the coefficient of z^n in the formal series in z:

$$\prod_{e \in E} (1 - z^{|e|})^{-1}.$$

This implies

$$(1 - kz)^{-1} = \prod_{e \in E} (1 - z^{|e|})^{-1}.$$

Or

$$(1 - kz)^{-1} = \prod_{n > 0} (1 - z^n)^{-\psi_k(n)}. \tag{5.3.17}$$

Taking the logarithm on both sides, we obtain:

$$k^n = \sum_{d \mid n} d\psi_k(d),$$

which is equivalent to Witt's formula (by the Möbius inversion formula).

As an application of the Poincaré theorem, let us consider a bisection (X, Y) of A^* and $Z = X \cup Y$. Let λ be the mapping of Z into $\mathfrak{L}(A)$ defined inductively by $\lambda(a) = a$ for $a \in A$ and

$$\lambda(z) = [\lambda(y), \lambda(x)] \tag{5.3.18}$$

if $z = yx$, $y \in Y$, $x \in X$.

The subalgebras of $K\langle A \rangle$ generated by X and Y are freely generated by X and Y since X and Y are codes. Hence the subalgebras of $K\langle A \rangle_L$ generated by X and Y may be denoted $\mathfrak{L}(X)$ and $\mathfrak{L}(Y)$. The sum of $\mathfrak{L}(X)$ and $\mathfrak{L}(Y)$ is direct since $X^* \cap Y^* = 1$. The mapping λ extends to a morphism of $\mathfrak{L}(X)$ and $\mathfrak{L}(X)$ into $\mathfrak{L}(A)$.

PROPOSITION 5.3.11. *The mapping λ is an isomorphism from $\mathfrak{L}(X) \oplus \mathfrak{L}(Y)$ onto $\mathfrak{L}(A)$.*

Proof. Let i be the canonical morphism of $\mathfrak{L}(Z)$ onto its quotient by the Lie ideal I generated by the elements

$$[y, x] - yx, \quad y \in Y, x \in X.$$

Then the restriction of i to the submodule $\mathfrak{L}(X) \oplus \mathfrak{L}(Y)$ is an isomorphism of the module $\mathfrak{L}(X) \oplus \mathfrak{L}(Y)$ onto the module $\mathfrak{L}(Z)/I$. In fact, it is obviously injective and to see that it is surjective, it is enough to remark that, by the Jacobi identity, the equalities

$$[[u, z]z'] = [u[zz']] - [z[u, z']] \tag{5.3.19}$$

for $u \in \mathfrak{L}_{n-1}(X), z \in X, z' \in Y$ (resp. $u \in \mathfrak{L}_{n-1}(Y), z \in Y, z' \in X$) prove by induction on n that any element of $[\mathfrak{L}_n(X), \mathfrak{L}_1(Y)]$ (resp. $[\mathfrak{L}_n(Y), \mathfrak{L}_1(X)]$) is congruent mod I to an element of $\mathfrak{L}(X) \oplus \mathfrak{L}(Y)$ of degree at most n; and therefore that any element of $\mathfrak{L}(Z)$ is congruent mod I to an element of $\mathfrak{L}(X) \oplus \mathfrak{L}(Y)$.

Hence, there exists a unique Lie algebra $\mathfrak{L}$ on the module $\mathfrak{L}(X) \oplus \mathfrak{L}(Y)$ whose product, denoted (,) extends the product of $\mathfrak{L}(X)$ and $\mathfrak{L}(Y)$ and satisfies

$$(y, x) = yx, \quad y \in Y, x \in X.$$

The algebra $\mathfrak{L}$ is generated by A since, for any $z \in Z - A$, the equality $z = (y, x)$ for $z = yx, y \in Y, x \in X$, shows by induction on $|z|$ that z belongs to the subalgebra of $\mathfrak{L}$ generated by A.

The mapping λ of Z into $\mathfrak{L}(A)$ extends to a morphism from $\mathfrak{L}(Z)$ into $\mathfrak{L}(A)$ whose kernel contains the ideal I. Therefore λ is a morphism from the Lie algebra $\mathfrak{L}$ into $\mathfrak{L}(A)$. Since it is the identity on A, it is by Corollary 5.3.8 an isomorphism from $\mathfrak{L}$ onto $\mathfrak{L}(A)$. ∎

Example 5.3.12. (Lazard's elimination method). Let (X, Y) be the bisection of A^*:

$$X = a^*(A - a), \quad Y = a,$$

(cf. Example 5.2.1). Then, by Proposition 5.3.11:

$$\mathcal{L}(A) = Ka + \mathcal{L}(B) \tag{5.3.20}$$

where B is the set of elements

$$[\underbrace{a[a\ldots[a}_{i},b]\ldots]]$$

for $i \geqslant 0$ and $b \neq a$. This process, known as Lazard's elimination method, may be iterated to obtain a basis of $\mathcal{L}(A)$ (see Bourbaki 1971).

As another application of the Poincaré theorem, the following result gives a useful characterization of the elements of $\mathcal{L}(A)$. Let $R = K\langle A\rangle$ and consider the algebra $R \otimes R$ that is the tensor product of the module R with itself equipped with the componentwise product:

$$(u \otimes v)(r \otimes s) = (ur \otimes vs). \tag{5.3.21}$$

In fact, $R \otimes R$ is isomorphic to the algebra of the monoid $A^* \times A^*$, the direct product of A^* and A^*.

The *diagonal mapping* δ of R is the unique morphism of algebras of R into $R \otimes R$ such that for $a \in A$,

$$\delta(a) = a \otimes 1 + 1 \otimes a. \tag{5.3.22}$$

It is useful to visualize δ as follows: the image of a word $w \in A^+$ under δ is the sum of all pairs of words $u, v \in A^*$ such that w is a shuffle of u and v (see Problem 5.3.5).

THEOREM 5.3.13. (Friedrichs'). *If the characteristic of K is 0, an element $u \in K\langle A\rangle$ is in $\mathcal{L}(A)$ iff:*

$$\delta(u) = u \otimes 1 + 1 \otimes u. \tag{5.3.23}$$

Proof. Let P be the set of elements of $R = K\langle A\rangle$ such that Eq. (5.3.23) holds. Then P contains A by (5.3.22) and it is a subalgebra of R_L since for $u, v \in P$,

$$\delta[u,v] = [\delta(u), \delta(v)] = [u \otimes 1 + 1 \otimes u, v \otimes 1 + 1 \otimes v]$$
$$= [u,v] \otimes 1 + 1 \otimes [u,v].$$

Therefore P contains $\mathcal{L}(A)$. Conversely, let $E = \{e_1, e_2, \ldots\}$ be a basis for $\mathcal{L}(A)$. Then, since R is the enveloping algebra of $\mathcal{L}(A)$, the elements

$e_1{}^{k_1}e_2{}^{k_2}\cdots e_m{}^{k_m}$ for $m \geqslant 0$, $k_i \geqslant 0$ form a basis of R and the elements

$$e_1{}^{k_1}e_2{}^{k_2}\cdots e_m{}^{k_m}\otimes e_1{}^{l_1}e_2{}^{l_2}\cdots e_m{}^{l_m}$$

form a basis for $R \otimes R$. We have

$$
\begin{aligned}
\delta\left(e_1{}^{k_1}e_2{}^{k_2}\cdots e_m{}^{k_m}\right) &= \left(e_1\otimes 1 + 1\otimes e_1\right)^{k_1}\cdots\left(e_m\otimes 1 + 1\otimes e_m\right)^{k_m}\\
&= e_1{}^{k_1}e_2{}^{k_2}\cdots e_m{}^{k_m}\otimes 1 + k_1 e_1{}^{k_1-1}e_2{}^{k_2}\cdots e_m{}^{k_m}\otimes e_1\\
&\quad + k_2 e_1{}^{k_1}e_2{}^{k_2-1}\cdots e_m{}^{k_m}\otimes e_2 + \cdots +\\
&\quad + k_m e_1{}^{k_1}e_2{}^{k_2}\cdots e_m{}^{k_m-1}\otimes e_m + \rho
\end{aligned}
\tag{5.3.24}
$$

where ρ belongs to the submodule of $R\otimes R$ generated by the elements of the form

$$e_1{}^{j_1}e_2{}^{j_2}\cdots e_m{}^{j_m}\otimes e_1{}^{l_1}e_2{}^{l_2}\cdots e_m{}^{l_m} \quad\text{with } \sum_{i=1}^{m} l_i \geqslant 2.$$

The second term through the $(m+1)$st term of the right-hand side of (5.3.24) do not belong to that submodule; therefore, in order that $\delta(u)$ be a linear combination of the basis elements of the form

$$e_1{}^{k_1}e_2{}^{k_2}\cdots e_m{}^{k_m}\otimes 1 \quad\text{and}\quad 1\otimes e_1{}^{j_1}e_2{}^{j_2}\cdots e_m{}^{j_m},$$

it is necessary that in the expression of u in terms of the chosen basis, only elements $e_1{}^{k_1}e_2{}^{k_2}\cdots e_m{}^{k_m}$ with one $k_i = 1$ and the other $k_j = 0$ occur with nonzero coefficients (since K is of characteristic 0). This means that u is a linear combination of the e_i and therefore that $u \in \mathfrak{L}(A)$. This proves that $P \subset \mathfrak{L}(A)$ and therefore concludes the proof. ∎

There is an equivalent form of Friedrichs' theorem stating that an element $\sigma \in K\langle A\rangle$ is in $\mathfrak{L}(A)$ iff it is orthogonal to all shuffles $u \circ v$ for $u, v \in A^+$ (see Problem 5.3.4).

We now introduce the logarithm and exponential functions as partial functions defined on the algebra $K\langle\langle A\rangle\rangle$ of formal series over A. We shall need to use these functions in more general algebras and we begin by some preliminaries.

From now on, let K be a field of characteristic 0.

Let M be a monoid that is a direct product of a finite number of free monoids and $S = K^M$ be the set of mappings of M into K. We denote as $\langle\sigma, m\rangle$ the value of $\sigma \in S$ for $m \in M$. A family $(\sigma_i)_{i\in I}$ of elements of S is said to be locally finite if the set

$$\{i \in I \,|\, \langle\sigma_i, \; m\rangle \neq 0\}$$

is finite. If $(\sigma_i)_{i \in I}$ is a locally finite family of elements of S the sum

$$\sigma = \sum_{i \in I} \sigma_i$$

is defined by

$$\langle \sigma, m \rangle = \sum_{i \in I} \langle \sigma_i, m \rangle.$$

In particular, for all $\sigma \in S$ we have

$$\sigma = \sum_{m \in M} \langle \sigma, m \rangle m$$

where we denote by $\langle \sigma, m \rangle m$ the element of S with value $\langle \sigma, m \rangle$ on m and 0 elsewhere.

If $\sigma, \tau \in S$, the product $\sigma \tau$ is defined by

$$\langle \sigma \tau, m \rangle = \sum_{m = uv} \langle \sigma, u \rangle \langle \tau, v \rangle. \qquad (5.3.25)$$

The hypothesis that M is a direct product of free monoids implies that the sum in (5.3.25) is finite. These operations turn S into a K-algebra.

When $M = A^*$, this algebra is the same as $K\langle\langle A \rangle\rangle$; when $M = A^* \times A^*$, the algebra S contains the algebra $K\langle A \rangle \otimes K\langle A \rangle$ used before; when M is the free commutative monoid over A, isomorphic with the direct product of the free monoids a^* for $a \in A$, then S is the same as $K[[A]]$.

Let $S^{(1)}$ be the ideal of S defined by

$$S^{(1)} = \{\sigma \in S \mid \langle \sigma, 1 \rangle = 0\}.$$

Then for any $\sigma \in S^{(1)}$, the family $(\sigma^n)_{n \in \mathbb{N}}$ is locally finite and therefore

$$\exp(\sigma) = 1 + \sigma + \frac{\sigma^2}{2!} + \frac{\sigma^3}{3!} + \cdots \qquad (5.3.26)$$

$$\log(1 + \sigma) = \sigma - \frac{\sigma^2}{2} + \frac{\sigma^3}{3} - \cdots \qquad (5.3.27)$$

are well-defined elements of S. A direct computation shows that

$$\exp(\log(1 + \sigma)) = 1 + \sigma, \log(\exp(\sigma)) = \sigma, \qquad (5.3.28)$$

so that exp and log are mutually inverse bijections of $S^{(1)}$ onto $1 + S^{(1)}$.

Moreover, if $\sigma_1, \sigma_2 \in S^{(1)}$ and $\sigma_1\sigma_2 = \sigma_2\sigma_1$, then:

$$\exp(\sigma_1)\exp(\sigma_2) = \exp(\sigma_1 + \sigma_2), \tag{5.3.29}$$

$$\log(1+\sigma_2)(1+\sigma_2) = \log(1+\sigma_1) + \log(1+\sigma_2), \tag{5.3.30}$$

$$\log(\exp\sigma_1 \exp\sigma_2) = \sigma_1 + \sigma_2. \tag{5.3.31}$$

Let M and N be two monoids that are direct products of free monoids, and $S = K^M, T = K^N$. A morphism α of M into T is said to be *continuous* if $\alpha(M-1) \subset T^{(1)}$. If α is continuous, the family $(\alpha(m))_{m\in M}$ is locally finite, and therefore α extends to a morphism of S into T by

$$\alpha(\sigma) = \sum_{m\in M} \langle \sigma, m\rangle \alpha(m).$$

We then have for any $\sigma \in S^{(1)}$,

$$\exp(\alpha(\sigma)) = \alpha(\exp(\sigma)), \tag{5.3.32}$$

$$\log(\alpha(1+\sigma)) = \alpha\log(1+\sigma). \tag{5.3.33}$$

We now turn back to the algebra $S = K\langle\langle A\rangle\rangle$; any element $\sigma \in S$ may be written uniquely

$$\sigma = \sum_{n\geq 0} \sigma_n$$

with σ_n a homogeneous polynomial of degree n. The element σ will be called a *Lie element* if for any $n \geq 0$, σ_n belongs to $\mathcal{L}_n(A)$. We denote by $\overline{\mathcal{L}}(A)$ the set of Lie elements in S.

THEOREM 5.3.14 (Campbell, Baker, Hausdorff). *For any $a, b \in A$, the element* $\log(\exp a \exp b)$ *is a Lie element.*

Proof. The morphism δ of $R = K\langle A\rangle$ into $R\otimes R$ extends to a morphism δ of $S = K\langle\langle A\rangle\rangle$ into $T = K^{A^*\times A^*}$ by

$$\delta(\sigma) = \sum_{w\in A^*} \langle \sigma, w\rangle \delta(w).$$

And, by Friedrichs' theorem, σ is a Lie element iff $\delta(\sigma) = 1\otimes\sigma + \sigma\otimes 1$; in fact, let

$$\sigma = \sum_{n\geq 0} \sigma_n$$

with σ_n homogeneous of degree n. Then, if $\sigma \in \bar{\mathcal{L}}(A)$,

$$\delta(\sigma) = \sum_{n \geq 0} \delta(\sigma_n) = \sum_{n \geq 0} (1 \otimes \sigma_n + \sigma_n \otimes 1) = 1 \otimes \sigma + \sigma \otimes 1.$$

Conversely, if $\delta(\sigma) = 1 \otimes \sigma + \sigma \otimes 1$, then for each $n \geq 0$, $\delta(\sigma_n) = 1 \otimes \sigma_n + \sigma_n \otimes 1$ and therefore $\sigma_n \in \mathcal{L}(A)$; hence $\sigma \in \bar{\mathcal{L}}(A)$.

Now for any $\sigma \in 1 + S^{(1)}$, $\log(\sigma) \in \bar{\mathcal{L}}(A)$ iff

$$\delta(\log(\sigma)) = (\log \sigma) \otimes 1 + 1 \otimes (\log \sigma). \tag{5.3.34}$$

But, since the morphism δ of A^* into T is continuous, we have by (5.3.33),

$$\delta(\log(\sigma)) = \log(\delta(\sigma));$$

and, also by (5.3.33),

$$(\log \sigma) \otimes 1 = \log(\sigma \otimes 1), \qquad 1 \otimes (\log \sigma) = \log(1 \otimes \sigma).$$

Hence, (5.3.34) is equivalent to

$$\log(\delta(\sigma)) = \log(\sigma \otimes 1) + \log(1 \otimes \sigma). \tag{5.3.35}$$

And as $(1 \otimes \sigma)(\sigma \otimes 1) = (\sigma \otimes 1)(1 \otimes \sigma) = \sigma \otimes \sigma$, (5.3.35) is equivalent to

$$\log(\delta(\sigma)) = \log(\sigma \otimes \sigma). \tag{5.3.36}$$

Taking the exponential of both members of (5.3.36), we obtain that $\log(\sigma) \in \bar{\mathcal{L}}(A)$ iff

$$\delta(\sigma) = \sigma \otimes \sigma. \tag{5.3.37}$$

This proves the theorem, since for $a, b \in A$,

$$\begin{aligned}
\delta(\exp a \exp b) &= \delta(\exp a)\delta(\exp b) \\
&= (\exp a \otimes \exp a)(\exp b \otimes \exp b) \\
&= \exp a \exp b \otimes \exp a \exp b
\end{aligned}$$

proving that $\log(\exp a \exp b) \in \bar{\mathcal{L}}(A)$.   ∎

The series $\log(\exp a \exp b)$ is called the Hausdorff series; it is easy to calculate its first terms:

$$\log(\exp a \exp b) = a + b + \tfrac{1}{2}[a, b] + \tfrac{1}{12}[[a, b]b] + \tfrac{1}{12}[a,[a, b]] + \cdots \tag{5.3.38}$$

Let us denote by S' the set of elements σ of S such that each of its homogeneous components σ_n belongs to the subspace of $R = K\langle A\rangle$ generated by the $[u, v]$ for $u \in A^i$, $v \in A^j$, $i + j = n$. In other words, $\sigma \in S'$ iff for each $n \geq 0$, $\sum_{i=0}^{n}\sigma_i$ belongs to the submodule

$$R' = [R, R],$$

which is by definition the submodule of R generated by the $[r, s]$ for $r, s \in R$. By definition, the set $\bar{\mathfrak{L}}(A)$ of Lie elements is included in $R_1 \oplus S'$.

COROLLARY 5.3.15. For any $\sigma, \tau \in 1 + S^{(1)}$, the element

$$z = \log \sigma\tau - \log \sigma - \log \tau$$

belongs to S'.

Proof. If A has just one element, then $\sigma\tau = \tau\sigma$ and $z = 0$. Else, let α be a continuous morphism of A^* into S such that $\alpha(a) = \log \sigma$, $\alpha(b) = \log \tau$ for some $a, b \in A$. Then $y = \log(\exp a \exp b) - a - b$ belongs to S' by Eq. (5.3.38). But Since α is continuous $\alpha(R') \subset R'$ and also $\alpha(S') \subset S'$. Hence $z = \alpha(y) \in S'$. ∎

It is important to realize that Corollary 5.3.15 is a much weaker statement than Theorem 5.3.14; for a direct proof, see Problem 5.3.7.

Corollary 5.3.15 may be generalized to infinite products as follows: Let I be a totally ordered set and $(\sigma_i)_{i \in I}$ be a locally finite family of elements of $S^{(1)}$; the infinite product

$$\sigma = \prod_{i \in I} (1 + \sigma_i)$$

is defined as

$$\sigma = \sum_{J} \left(\prod_{i \in J} \sigma_i \right)$$

where J runs over the set of finite subsets of I. We then have:

COROLLARY 5.3.16. If $(\sigma_i)_{i \in I}$ is a locally finite family of elements of $1 + S^{(1)}$, then the element

$$z = \log \prod_{i \in I} \sigma_i - \sum_{i \in I} \log \sigma_i$$

belongs to S'.

5.4. The Theorem of Factorizations

Recall that a family $(X_i)_{i \in I}$ of subsets of A^+ indexed by a totally ordered set I is a factorization of the free monoid A^* if any word $w \in A^+$ may be written uniquely as:

$$w = x_1 x_2 \cdots x_n, \quad \text{with} \quad n \geqslant 1, x_i \in X_{j_i}, j_1 \geqslant j_2 \geqslant \cdots \geqslant j_n. \quad (5.4.1)$$

It was shown in Section 5.1 that the family $(l)_{l \in L}$ of Lyndon words is a factorization and in Section 5.2, we studied the factorizations indexed by a two-element set, called bisections.

The main result of this section is the following theorem

THEOREM 5.4.1 (Schützenberger). *Let* $(X_i)_{i \in I}$ *be a family of subsets of* A^+ *indexed by a totally ordered set* I. *This family is a factorization of* A^* *iff two of the following conditions are satisfied*:

   (i) *each word* $w \in A^+$ *admits at least one factorization* (5.4.1).
   (ii) *each word* $w \in A^+$ *admits at most one factorization* (5.4.1).
   (iii) *Each class* C *of conjugate elements of* A^+ *meets one and only one of the submonoids* $M_i = X_i^*$, $(i \in I)$ *whose minimal generating set is* X_i, *and the elements of* $C \cap M_i$ *are conjugate within* M_i.

Proof. First suppose that the family $(X_i)_{i \in I}$ is a factorization of A^*; that is, that it satisfies conditions (i) and (ii). Then, using characteristic series with coefficients in the field $\mathbb{Q}$ of rational numbers, we have:

$$\mathbf{A}^* = \prod_{i \in I} (\mathbf{X}_i)^*; \quad (5.4.2)$$

and this may be written

$$(1 - \mathbf{A})^{-1} = \prod_{i \in I} (1 - \mathbf{X}_i)^{-1} \quad (5.4.3)$$

Taking the logarithm of both members, we obtain by Corollary 5.3.16,

$$\log(1 - \mathbf{A})^{-1} = \sum_{i \in I} \log(1 - \mathbf{X}_i)^{-1} + z, \quad (5.4.4)$$

where z is a series each homogeneous component of which belongs to the space generated by the elements $[u, v] = uv - vu$, for $u, v \in A^+$.

Let then C be a class of conjugate elements of A^*. Let n be their common length and p their exponent; that is, the elements of C are the n/p distinct conjugates of a word u^p, with $|u| = n/p$ (see Proposition 1.3.3.). We then have:

$$\langle \log(1-\mathbf{A})^{-1}, \mathbf{C} \rangle = \langle \sum_{i>0} \frac{1}{i} \mathbf{A}^i, \mathbf{C} \rangle = \frac{1}{n} \langle \mathbf{A}^n, \mathbf{C} \rangle = \frac{1}{p}, \qquad (5.4.5)$$

since C contains n/p distinct elements. Now, using Eq. (5.4.4) and $\langle z, \mathbf{C} \rangle = 0$, we obtain

$$\sum_{i \in I} \langle \log(1-\mathbf{X}_i)^{-1}, \ \mathbf{C} \rangle = \frac{1}{p}. \qquad (5.4.6)$$

But for each $i \in I$, we have

$$\log(1-\mathbf{X}_i)^{-1} = \sum_{j>0} \frac{1}{j} \mathbf{X}_i^j \qquad (5.4.7)$$

and therefore, since each X_i is a code:

$$\langle \log(1-\mathbf{X}_i)^{-1}, \mathbf{C} \rangle = \sum_{j>0} \frac{1}{j} \operatorname{Card}(\mathbf{X}_i^j \cap \mathbf{C}). \qquad (5.4.8)$$

But if $w \in X_i^j \cap C$, it has at least j/p, distinct conjugates in $X_i^j \cap C$ and therefore, for each $i \in I$ such that $X_i^* \cap C \neq \varnothing$, we have

$$\langle \log(1-\mathbf{X}_i)^{-1}, \ \mathbf{C} \rangle \geq \frac{1}{p} \qquad (5.4.9)$$

with equality iff all elements of $X_i^* \cap C$ are conjugate within X_i^*. Comparing with (5.4.6), we deduce from (5.4.9) that C intersects exactly one submonoid X_i^*, proving that condition (iii) holds.

Conversely, suppose that condition (iii) holds. We first prove that each X_i is a code; let in fact $X = X_i$ and suppose that

$$x_1 x_2 \cdots x_n = y_1 y_2 \cdots y_m$$

with $n, m \geq 1$, $x_j, y_k \in X$; we may suppose that $|x_1| \geq |y_1|$ and let $x_1 = y_1 u$. Then, denoting $v = x_2 \cdots x_n y_1$, we have

$$uv, vu \in X^*.$$

Let $uv = t^p$, with t primitive. By condition (iii), t has a conjugate t' in one of the X_j^* and then $t'^p \in X_j^*$, $t^p \in X^*$ forces $X_j = X$, hence $t \in X^*$ since t^p and t'^p are conjugate. Let $u = t^k r$, $v = st^l$ with $t = rs$, $k + l + 1 = p$. Then

$vu = (sr)^p$ and again $sr \in X^*$; by condition (iii), rs and sr have to be conjugate within X^* but since $t = rs$ is primitive, this forces, $r, s \in X^*$. Hence, $u, v \in X^*$; since $x_1 = y_1 u$ and X is the minimal generating set of X^*, this implies $u = 1$ and $x_1 = y_1$. This proves by induction on $n + m$ that X is a code.

Therefore, if condition (iii) holds, equalities (5.4.7), (5.4.6), and therefore (5.4.4) also hold. Let now α be the canonical morphism of the ring $Q\langle\langle A \rangle\rangle$ onto the ring $Q[[A]]$ of series in the commuting variables of A. Since $\alpha(z) = 0$, we have

$$\alpha \log(1 - \mathbf{A})^{-1} = \sum_{i \in I} \alpha \log(I - \mathbf{X}_i)^{-1}$$

and, by formula (5.3.33), this implies

$$\log(1 - \alpha \mathbf{A})^{-1} = \sum_{i \in I} \log(1 - \alpha \mathbf{X}_i)^{-1}$$

or

$$(1 - \alpha \mathbf{A})^{-1} = \prod_{i \in I} (1 - \alpha \mathbf{X}_i)^{-1}. \qquad (5.4.10)$$

Now if we set

$$R = \mathbf{A}^* - \prod_{i \in I} \mathbf{X}_i^*,$$

condition (i) means that the coefficients of $-R$ are nonnegative, while (ii) expresses that the coefficients of R are nonnegative. And, by (5.4.10), condition (iii) implies that $\alpha(R) = 0$ Therefore, if condition (iii) is satisfied, conditions (i) and (ii) are equivalent since both imply that $R = 0$. This concludes the proof of the theorem. ∎

In the case of a bisection (X, Y), the fact that condition (iii) of Theorem 5.4.1 is satisfied may be directly verified as follows: First, as a consequence of (5.2.5) we have

$$Y^* X^* \subset X^* \cup Y^* \qquad (5.4.11)$$

and therefore each word $w = xy$ has a conjugate in X^* or Y^*, namely yx. Further, if $uv \in X^*$ and $vu \in Y^*$, then by (5.2.7) and (5.2.8), we have $v \in X^* \cap Y^*$ and therefore $v = 1$; hence each class of conjugate elements intersects exactly one of the two submonoids X^* and Y^*. Finally, if two conjugate elements uv, vu are in X^*, then by (5.2.7) $u, v \in X^*$ and they are therefore conjugate within X^*.

In the case of Lyndon words, condition (iii) was used in the definition. Notice that, knowing Theorem 5.4.1, it is enough to prove that one of the two conditions (i) or (ii) holds to prove that the family $(l)_{l \in L}$ is a factorization of A^*.

Call a factorization $(X_i)_{i \in I}$ of A^* *complete* if all the sets $X_i, i \in I$ are singleton sets.

COROLLARY 5.4.2. *If $(x_i)_{i \in I}$ is a complete factorization of A^*, the $x_i, i \in I$ are a system of representatives of the conjugacy classes of primitive words.*

This section ends with further examples of factorizations (others are treated in the exercises).

Let $\varphi: A^* \to \mathbb{R}$ be a morphism from the free monoid A^* into the additive monoid $\mathbb{R}$ and for every $r \in \mathbb{R}$, let

$$C_r = \{v \in A^+ \mid \varphi(v) = r|v|\},$$

$$B_r = C_r - \left(\bigcup_{s \geq r} C_s\right) A^+.$$

The following result is due to Spitzer (1956).

THEOREM 5.4.3. *The family $(B_r)_{r \in \mathbb{R}}$ (with the usual ordering on $\mathbb{R}$) is a factorization of A^*.*

The proof is left to the reader as an exercise (Problem 5.4.6). Observe that this result has a very simple graphical interpretation: to any word $w = a_1 a_2 \cdots a_n$, associate the set of points $(i, \varphi(a_1 \cdots a_i))$ in the plane, with $0 \leq i \leq n$. The convex hull of the set formed by these points induces the factorization $w = v_1 \cdots v_m$ by taking the points that belong to the hull. As an example, with $\varphi(a) = +1$, $\varphi(b) = -1$, $w = baababbab$, we have $w = (baa)(ba)(bba)(b) \in B_{1/3} B_0 B_{-1/3} B_{-1}$ (see Figure 5.3).

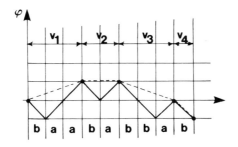

Figure 5.3. Spitzer's factorization.

Our next example is the family of *Viennot factorizations* defined in the following result:

THEOREM 5.4.4. *Let X be a totally ordered subset of A^+ such that*:

(i) $X = A \cup \{xy \mid\ x, y \in X,\ x < y\}$
(ii) *for any $x, y \in X$ such that $x < y$, then $x < xy < y$.*
     Then the family $(x)_{x \in X}$ is a factorization of A^*.

Proof. We prove by induction on the length of $w \in A^+$ that for any alphabet A and $X \subset A^+$ satisfying conditions (i) and (ii), w may be written uniquely

$$w = x_1 x_2 \cdots x_n, \tag{5.4.12}$$

with $n \geqslant 1$ and with $x_i \in X$ and $x_1 \geqslant x_2 \geqslant \cdots \geqslant x_n$. The property is true for $w \in A$; let $w \in A^+$ be such that $|w| \geqslant 2$. First we may assume that each letter of A has an occurrence in w — that is, that $A = \mathrm{alph}(w)$. If not, we would define $A' = \mathrm{alph}(w)$ and $X' = X \cap A'^*$, which has the same properties as the set X.

As A is finite, it has a minimal element, say $a \in A$; then a is also the minimal element of X since any $z \in X - A$ may be written $z = xy$, with $x, y \in X$ and $x < y$ by condition (i) and that, by (ii), $x < xy$, proving by induction that $a < xy$.

Let

$$Z = a^*(A - a)$$

Then $Z \subset X$ since it is easy to prove by induction on $i \geqslant 0$ that $a^i b \in X$ for any $b \in A - a$.

Further, any $x \in X - Z - a$ may be written $x = x'x''$ with $x', x'' \in X - a, x' < x''$. We prove this by induction on the length of x: let the pair $(x', x'') \in X$ with $x = x'x''$, $x' < x''$, be chosen with $|x'|$ maximal. If we had $x' = a$, then first $x'' \neq a$ and $x' \notin Z$ since otherwise $x \in Z$; thus, by induction, x'' may be written $x'' = t't''$ with $t', t'' \in X - a, t' < t''$. Therefore $a < t'$; but this implies $at' \in X$, by (i); and by (ii):

$$at' < t' < t't'' < t'',$$

and this contradicts the definition of x' since $|x'| < |at'|$ and $at', t'' \in X, at' < t'', x = at't''$.

As a consequence, we have

$$X \subset Z^* \cup a. \tag{5.4.13}$$

Let

$$\beta: B \to Z$$

be a bijection from a set B onto Z extended to an isomorphism from B^* to Z^*. Then, the set $Y = \beta^{-1}(X-a)$ ordered by

$$y < y' \quad \text{if} \quad \beta(y) < \beta(y')$$

satisfies conditions (i) and (ii) (with B instead of A). For condition (i)

$$Y = B \cup \{y'y'' \mid y', y'' \in Y, \quad y' < y''\},$$

we clearly have the right member included in the first. Conversely, let $y \in Y - B$ and $x = \beta(y)$; as $x \in X - Z - a$, we proved above the existence of a pair $x', x'' \in X - a$ such that $x = x'x''$, $x' < x''$. Then $y = y'y''$ with $y' = \beta^{-1}(x')$, $y'' = \beta^{-1}(x'')$ and $y' < y''$, proving that condition (i) holds for Y; condition (ii) is obviously satisfied.

Coming back to the word $w \in A^+$; it can be written in a unique way:

$$w = za^i,$$

with $z \in Z^*$, $i \geq 0$ (that is, the pair (Z, a) is a bisection of A^*). And for any factorization (5.4.12) of w:

$$z = x_1 x_2 \cdots x_{n-i}, \qquad x_{n-i+1} = \cdots = x_n = a,$$

since $X \subset Z^* \cup a$ by (5.4.13) and a is the minimal element of X. But

$$|\beta^{-1}(z)|_B = |w|_{A-a},$$

which is strictly less than $|w|$ since $a \in \mathrm{alph}(w)$ and $|w| \geq 2$. Therefore, by induction, the word $u = \beta^{-1}(z)$ may be written uniquely:

$$u = y_1 y_2 \cdots y_m,$$

with $m \geq 0$, $y_i \in Y$, $y_1 \geq y_2 \geq \cdots \geq y_m$. This forces $m = n - i$ and $x_1 = \beta(y_1), \ldots, x_{n-i} = \beta(y_{n-i})$ proving both the existence and unicity of the factorization (5.4.12) $\blacksquare$

The family of Viennot factorizations contains the factorization in Lyndon words. In fact, the set L of Lyndon words satisfies condition (i) by Proposition 5.1.3 and condition (ii) by Proposition 5.1.2.

It can be shown that Spitzer's factorization may be refined so as to give a Viennot factorization (Viennot 1978); the factorization obtained has been considered by Foata (1965). Viennot factorizations are also closely related to bisections; it may be observed that the proof of Theorem 5.4.4 uses a sequence of auxiliary bisections to obtain the factorization as a refinement.

Notes

Lyndon words were introduced under the name of *standard lexicographic sequences* in (Lyndon 1954) and (Lyndon 1955). They were used to construct a basis of the free abelian group F_n/F_{n+1}, where F_n is nth derived group of the free group F on a set A; in fact F_n/F_{n+1} is isomorphic to $\mathcal{L}_n(A)$, the nth homogeneous component of the free Lie algebra via the Magnus transformation (see Probelm 5.3.9); the construction of the basis of F_n/F_{n+1} is in fact the same as the construction of the basis of $\mathcal{L}(A)$ associated with Lyndon words described in Section 5.3. The properties of Lyndon words presented in section 5.1 are from (Chen, Fox, and Lyndon 1958). Other bases of $\mathcal{L}(A)$ have been constructed. One is that of P. Hall using the so-called *basic commutators* (see M. Hall 1959 and Problem 5.4.3). It was later generalized in Meier-Wunderli 1952. In Shirshov 1958 a basis for $\mathcal{L}(A)$ is defined that is, up to symmetries, identical to that of Lyndon.

The notion of a bisection, like the general study of factorizations of free monoids, goes back to Schützenberger 1965; this paper contains Proposition 5.2.1. and Theorem 5.4.1. In a previous paper (Schützenberger 1959), the relation between factorizations of free monoids and basis of free Lie algebras is studied.

Factorizations of free monoids have been extensively studied by Viennot. Proposition 5.3.9 and Theorem 5.4.4. appear in Viennot 1978. Many of his results are not presented here. One of them is the construction of trisections —that is, factorizations of the type $A^* = X^* Y^* Z^*$ (Viennot 1974). Another is the characterization, in terms of the construction of bisections given by Corollary 5.2.2, of those bisections for which the sets X and Y are recognizable, in the sense of Eilenberg 1974, (Viennot 1974). The factorizations corresponding to Hall sets (see Problem 5.4.3) are also proved to be exactly those that can be "locally" obtained by iterating bisections (Lazard's factorizations in Viennot 1972), therefore giving a basis of the free Lie algebra $\mathcal{L}(A)$; part of this result had been proved in (Michel 1974).

The connections of factorizations with combinatorial probabilities (especially fluctuations of random variables) where noted in (Schützenberger 1965). They are based on the work of Sparre-Andersen and Spitzer (see Spitzer 1956). The idea of the construction in Proposition 5.2.4 is credited in (Foata and Schützenberger 1971) to Richards and to Farrell (Farrell 1965).

This chapter's treatment of free Lie algebras follows essentially Jacobson 1962. For further results and references, see also Bourbaki 1971, 1972 and Magnus, Karass, and Solitar 1976. The connections of the free Lie algebra with the shuffle product appears in Chen, Fox, and Lyndon 1958 and Ree 1958 (see Problems 5.3.2–5.3.6).

Efficient computations in free Lie algebras have been considered by Michel (1974), who computed tables of the coefficients of the Hausdorff series. It has been pointed out by Viennot (1978) that the basis of $\mathcal{L}(A)$ associated to Lyndon words was especially convenient for practical uses since, by Lemma 5.3.2, it is possible to restrict the development of a Lie element u to the Lyndon monomials and then to invert triangular matrices to obtain the development of u in the basis $\lambda(L)$. The complexity of factorizing a word into Lyndon words has been considered by Duval (1980), who has shown that it is possible to do this using only a number of comparisons of letters that is a linear function of the length of the word.

Problems

Section 5.1

5.1.1. Prove the following property of lexicographic order:

$$u < w < uv \Leftrightarrow w = ut, t < v.$$

5.1.2. Show that if w is longer than three times its period p, there exists a factorization of w as $w = urrt$ such that r has period p. (This is a particular case of the critical factorization theorem; see Chapter 8) (*Hint*: Let $w = (v)^n v'$ with $|v| = p$ and v' a left factor of v; then define r as the Lyndon word conjugate to v.)

**5.1.3. Show that $w \in A^+$ is a Lyndon word iff for any nontrivial factorization $w = uv$, there exists a shuffle of u and v greater than w (Chen, Fox, and Lyndon 1958).

5.1.4. Let $\mathrm{Card}(A) = k$ and define $\alpha(k, n)$ as the minimum number of words in a set $S \subset A^n$ satisfying $\mathrm{Card}(A^* - A^* S A^*) < +\infty$; that is, all but a finite number of words have a factor in S.

a. Show that $\alpha(1, n) = 1, \alpha(k, 1) = k, \alpha(k, 2) = k(k+1)/2, \alpha(2, 5) = 9$.

b. Show that $\alpha(k, n) \geq k^n / n$.
(*Hint*: Use the fact that the set S must contain an element from each conjugate class of words of length n.)

c. Define a set $T_n \subset A^*$ as the left factors of length n of the powers of Lyndon words. Show that $\mathrm{Card}(A^* - A^* T_n A^*)$ is finite.

d. Deduce from (c) that for each $\varepsilon > 0$, there exists an integer $k(\varepsilon)$ such that $nk^{-n}\alpha(k, n) < 1 + \varepsilon$ for all n and $k \geq k(\varepsilon)$.

e. Define a set $S_n \subset A^*$ consisting of all the words a^n, $a \in A$ and the words of the form at, where $a \in A$, $t \in T_{n-1}$, and $t < a$. Show that $\mathrm{Card}(A^* - A^* S_n A^*)$ is finite.

f. Deduce from (e) that for each $\varepsilon > 0$, there exists an integer $n(k, \varepsilon)$ such that $nk^{-n}\alpha(k, n) < 1 + \varepsilon$, for $n \geqslant n(k, \varepsilon)$.

g. Conclude, by (d) and (f), that if $\max(k, n)$ tends to infinity

$$\lim nk^{-n}\alpha(k, n) = 1.$$

(See Schützenberger 1964.)

5.1.5. Call standard factorization of a word $w \in A^2 A^*$ the pair $\sigma(w) = (l, m)$ such that $w = lm$ and m is the longest proper right factor of w in L. Show that if $n \in L$ is such that $n \leqslant m$, then $\sigma(lmn) = (lm, n)$.

5.1.6. For any $w \in L - \underline{A}$, let $w = mn$ with m the longest proper left factor of w in L. Prove that $n \in L$ and $m < n$.

5.1.7. For any $l \in L - A$, let $k = au$ with $a \in A, u \in A^+$; let n be the smallest right factor of u in L; prove that

$$\sigma(l) = (m, n),$$

with $l = mn$.

Section 5.2

5.2.1. Show directly (without using formal series) that if (X, Y) is a bisection of A^*, then:

a. $A \subset X \cup Y$,

b. No word of $X \cup Y$ has a proper left (resp. right) factor in X (resp. Y).

c. $YX \subset X \cup Y$

d. $Y^* X^* \subset X^* \cup Y^*$

e. each word in $X \cup Y - A$ may be written uniquely as a product $yx, y \in Y, x \in X$.

5.2.2. Let P and Q be two submonoids of A^* such that $P \cap Q = 1$; define the left and right associates of P and Q, respectively, to be

$$X^* = \{ x \in X^* | x = uv \Rightarrow v \in P \},$$
$$Y^* = \{ y \in Y^* | y = uv \Rightarrow u \in P \},$$

where X and Y are minimal generating sets of X^* and Y^*. Show that (X, Y) is then a bisection of A^* (compare with Example 5.2.5).

5.2.3. Let $D \subset \{a, b\}^*$ be the set defined in Example 5.2.2; let σ be morphism of A^* into $\mathbb{Z}$ defined by

$$\sigma(a) = -1, \qquad \sigma(b) = 1.$$

    a. Show that $w \in D$ iff $\sigma(w) = 0$ and $\sigma(u) < 0$ for any nonempty proper left factor u of w.

    b. Deduce from (a) that

$$D = a\mathbf{D}^*b.$$

Section 5.3

5.3.1. (Dynkin's theorem). Let K be a field and $R^{(1)}$ be the subspace of $R = K\langle A \rangle$ formed by polynomials with zero constant term. Let λ be the linear mapping of $R^{(1)}$ into $\mathcal{L}(A)$ defined inductively by $\lambda(a) = a$ for $a \in A$, and

$$\lambda(wa) = [\lambda(w), a], \quad \text{for} \quad w \in A^+, a \in A.$$

    a. Show that, for $u \in R^{(1)}$ and $v \in R$,

$$\lambda(uv) = \lambda(u)\theta(v)$$

where θ is the unique algebra morphism of R into End $\mathcal{L}$ such that

$$u\theta(v) = [u, v]$$

for $u, v \in \mathcal{L}(A)$.

    b. Deduce from (a) that

$$\lambda([u, v]) = [\lambda(u), v] + [u, \lambda(v)]$$

for $u, v \in \mathcal{L}(A)$.

    c. Let K be a field of characteristic 0; show that a homogeneous element $u \in R$ of degree $n > 0$ is in $\mathcal{L}(A)$ iff $\lambda(u) = nu$.

5.3.2. Let ρ be the linear mapping of $R^{(1)}$ into $R^{(1)}$ defined inductively by $\rho(a) = a$ if $a \in A$ and

$$\rho(aub) = \rho(au)b - \rho(ub)a, \quad \text{if} \quad a, b \in A, \quad u \in A^*.$$

Show that ρ is the adjoint mapping of the mapping λ of Problem 5.3.4, that is, that for all $u, v \in R^{(1)}$

$$\langle \rho(u), v \rangle = \langle u, \lambda(v) \rangle$$

where $\langle , \rangle$ is the standard scalar product on R defined by $\langle u, v \rangle = 0$ or 1, for $u, v \in A^*$, according to $u \neq v$ or $u = v$. (see Ree 1958.)

*5.3.3. Let $\circ$ be the shuffle product on R (cf. Chapter 6) defined inductively by

$$ua \circ vb = (u \circ vb)a + (ua \circ v)b$$

for $u, v \in A^*, a, b \in A$. Show that for any $w \in A^n, n > 0$,

$$\sum_{\substack{w = uv \\ v \neq 1}} u \circ \rho(v) = nw,$$

(See Ree 1958.)

5.3.4. Let K be a field of characteristic 0; deduce from 5.3.3 that an element $u \in R^{(1)}$ is in $\mathcal{L}(A)$ iff

$$\langle v \circ w, u \rangle = 0$$

for any $v, w \in R^{(1)}$.

5.3.5. Use 5.3.4 to prove directly Friedrich's Theorem (Theorem 5.3.10).

*5.3.6. Let S be the (commutative) algebra over the module $R^{(1)}$ equipped with the shuffle product and L be the set of Lyndon words over the alphabet A. Show that if K is a field of characteristic 0, S is isomorphic to the symmetric algebra over the vector space $K[L]$ generated by L—that is, that any word $w \in A^+$ may be expressed uniquely as a linear combination of shuffles of Lyndon words. (Perrin, Viennot, 1981)

5.3.7. Let C be a class of conjugate words of $\{a, b\}^n$ and p be its exponent (that is, any $w \in C$ is a pth power of a primitive word). Let δ be the number of factors equal to ab in the circular word associated to C; more precisely $S = 0$ if $C \subset a^* \cup b^*$; if not δ is the number of factors ab in any word of $C \cap a(a, b)^* b$. Prove that the following equality holds for each $m \geq 1$:

$$\sum_{w \in C} \langle (a + b + ab)^m, w \rangle = \frac{m}{p} \binom{\delta}{n - m},$$

where $\binom{p}{q}$ is the binomial coefficient $\dfrac{p!}{q!(p - q)!}$.

5.3.8. Deduce from Problem 5.3.7 a direct proof of Corollary 5.3.15.

5.3.9. Let F be the free group over the set A and consider the Magnus transformation (cf. Chapter 6), which is the isomorphism of F into $\mathbb{Z}\langle\langle A \rangle\rangle$ given by

$$\mu(a) = 1 + a, \quad a \in A.$$

Let $(F_n)_{n \geq 0}$ be the lower central series of F defined by $F_0 = F$ and for $n \geq 1$,

$$F_n = [F_{n-1}, F],$$

where, for any two subgroups G, H of F, the symbol $[G, H]$ denotes the subgroup of F generated by the elements $xyx^{-1}y^{-1}$ for $x \in G$, $y \in H$. Let $\bar{\mu}$ be the mapping of F into $\mathbb{Z}\langle A \rangle$ associating to $x \in F$ the homogeneous component of lowest degree of $\mu(x) - 1$. Show that $\bar{\mu}$ induces an isomorphism of the group F_n / F_{n+1} onto $\mathfrak{L}_n(A)$. (See Magnus, Karrass, and Solitar 1976.)

Section 5.4

5.4.1. A submonoid M of A^* is, by definition *very pure* if

$$uv, vu \in M \Rightarrow u, v \in M.$$

a. Show that a very pure submonoid is free.
b. Show that a submonoid M of A^* is very pure iff the restriction to M of the conjugation in A^* is equal to the conjugation in M and for any $x \in A^*$, $n \geqslant 1$, $x^n \in M$ implies $x \in M$.

5.4.2. Let $A = \{1, 2, \ldots, n\}$ and, for $j \in A$, let X_j be the subset of A^+

$$X_j = j\{j+1, \ldots, n\}^*.$$

Prove that the family $(X_j)_{1 \leqslant j \leqslant n}$ is a factorization of A^*. (This is, with the reversed order, the factorization of Lemma 10.2.1.)

**5.4.3. *A magma M is a set with a binary operation denoted (u, v) for $u, v \in M$. Denote as $A^{(\,)}$ the free magma over A. A subset $H \subset A^{(\,)}$ is a *Hall set* if it is totally ordered and if the two following conditions are satisfied:

(i) $H = A \cup \{(a, u) \mid a \in A, \ u \in H, a < u\} \cup \{((u, v), w) \mid u, v, w \in H, (u, v) < w \leqslant v\}$.
(ii) $u, v \in H, (u, v) \in H \Rightarrow (u, v) < v$.

Let δ be the canonical mapping of $A^{(\,)}$ onto A^*; show that for any Hall set H, the family $(x)_{x \in \delta(H)}$ is a factorization of A^* (the order on $\delta(H)$ being induced by that on H). (See Viennot 1978.)

**5.4.4. Let $(x)_{x \in X}$ be a Viennot factorization. Let Π be the mapping of X into $A^{(\,)}$ defined inductively by $\Pi(a) = a$ for $a \in A$ and

$$\Pi(x) = (\Pi(y), \Pi(z))$$

if z is the longest proper right factor of x in X. Show that $\Pi(X)$ is a Hall set. (See Viennot 1978.)

**5.4.5. A totally ordered set $X \subset A^+$ is a *Lazard factorization* if for each integer $n > 0$, the set $X \cap A^{(n)} = \{x_1, x_2, \ldots, x_{k+1}\}$ of elements of X of length at most n ordered as $x_1 < x_2 < \cdots < x_{k+1}$ by the order of

X satisfies the two following conditions:

(i) $x_{k+1} \in Y_0 = A$, $x_k \in Y_1 = (Y_0 - x_{k+1})(x_{k+1})^*, \ldots, x_1 \in Y_k = (Y_{k-1} - x_2)(x_2)^*$,

(ii) $Y_k \cap A^{(n)} = x_1$.

a. Show that a Lazard factorization is a factorization of A^*.

b. Show that for any Hall set H, $\delta(H)$ is a Lazard factorization.

c. Show that for any Lazard factorization X, there exists a Hall set H such that $\delta(H) = X$. (See Viennot 1978.)

5.4.6. Prove Theorem 5.4.3.

CHAPTER 6

Subwords

6.0. Introduction

Let us recall the definition: a word f in A^* is a finite *sequence* of elements of A, called letters. We shall call a *subword* of a word f any sequence contained in the sequence f. The word *aba* for instance is a subword of the word *bacbcab* as well as of the word *aabbaa*. It can be observed immediately that two sub-sequences of f, distinct as subsequences, may define the same subword: thus *aba* is a subword of *bacbcab* in only one way but may be obtained as a subword of *aabbaa* in eight different ways.

A word f being given it is easy to compute the set of its subwords and their multiplicity; this computation is obtained by a simple induction formula. The main problem of interest in this chapter, sometimes implicitly but more often explicitly, is the one of the inverse correspondence. Under what conditions is a given set of words S the set of subwords, or a subset of certain kind of the set of subwords, of a word f? Once these conditions are met, what are the words f that are thus determined? In which cases are they uniquely determined? Some of these conditions on that set S are rather obvious. For instance if g is a subword of f, then any subword of g is a subword of f. Some conditions are more subtle; if for instance a and b are two letters of A, and if *ab* and *ba* are subwords of f, then at least one of the two words *aba* and *bab* is also a subword of f.

In Section 6.3 we shall consider the subwords with their multiplicity. It is possible to give a complete set of equations that express those relations. In Section 6.2 we shall be interested in the set of subwords of a word, without taking into account their multiplicity. More precisely we shall be concerned with the set T of words that have the same set of subwords of a given length m. It will be shown that if two words f and g are in T, there exists a word h also in T such that f and g are both subwords of h (Theorem 6.2.6) and that for any k less than $2m$, T has at most one word of length k (Theorem 6.2.16). Before these two main sections we shall recall in Section 6.1 the simple but basic result of Higman that in any infinite set of words there always exists one word that is a subword of another word in that set.

6.1.   The Division Ordering

If a word g is a subword of a word f we shall say that g *divides* f, and we shall denote it by $g \mid f$. If g does not divide f, we denote it by $g \nmid f$.

Division is a reflexive and transitive relation. The next remark is very simple as well: If g divides f then the length of g is less than or equal to the length of f, and if equality holds for the lengths, then g equals f. This argument will be used throughout the chapter and referred to as *the length argument*. As its first consequence, division is an antisymmetric relation. Finally note that if g divides f, and g' divides f', then gg' divides ff'. This paragraph is summed up by the following statement:

PROPOSITION 6.1.1. *Division is a partial ordering of A^*, which is compatible with the product of A^*.*

Figure 6.1 shows that A^*, equipped with the division ordering, is neither an inf- nor a sup-semilattice. The main property of division is given by:

THEOREM 6.1.2. *Any subset of words over a finite alphabet that are not comparable pairwise for the division ordering is finite.*

Since by the length argument there exists no infinite strictly descending chain of words, Theorem 6.1.2 yields that division is a *well partial ordering*, say by definition of a well partial ordering (see below). This result is due to Higman (1952), who gives it explicitly as a corollary of a much stronger statement. The direct proof given here is taken from Conway 1971 (pp. 62–63). The main idea in that proof is due to Nash-Williams (1963).

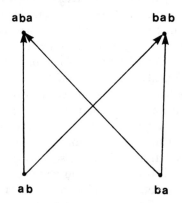

Figure 6.1. The relative situation of four words for the division ordering.

Proof. Suppose, ab absurdo, that there exist infinite sets of pairwise incomparable elements, and thus, in particular, that there exist infinite *division-free* sequences of words $f_1, f_2, \ldots, f_m, \ldots$ that is, sequences for which $i < j$ implies $f_i \nmid f_j$. Using the axiom of choice we can select an "earliest" such sequence, that is, one that satisfies conditions:

> f_1 is a shortest word beginning an infinite division-free sequence.
> f_2 is a shortest word such that f_1, f_2 begins such a sequence.
> f_3 is shortest such that f_1, f_2, f_3 begins such a sequence, and so on.

Since A is supposed to be finite, there exist infinitely many f_i that begin with the same letter, say $f_{i_1} = ag_1, f_{i_2} = ag_2, \ldots$ with $i_1 < i_2 < \cdots$ We then have the infinite division-free sequence $f_1, f_2, \ldots, f_{i_1-1}, g_1, g_2, \ldots$ which is "earlier" than the one we chose; this is a contradiction.   ∎

Observe that even though every set of pairwise incomparable elements is finite, the number of its elements is not bounded; indeed there exist arbitrarily large sets of pairwise incomparable elements (see Problem 6.1.4).

As Conway noted, Higman's theorem is little known and Theorem 6.1.2 has been rediscovered a number of times. The result is often credited to Haines (1969) (see also Problem 6.1.3). Kruskal (1972) gave a fairly complete history as well as a survey of the rich and well-developed theory of well quasi-ordering, in which Theorem 6.1.2 appears to be the most elementary result.

Well partial ordering may be defined by several equivalent properties that are worth mentioning, for they apply to our case. Recall first that a subset X of any ordered set E is an *ideal* if x in X and $x \leqslant y$ imply that y belongs to X. The *ideal generated* by X, denoted by $\bar{X}$, is the smallest ideal of E containing X, and is equal to the set of elements of E greater than at least one element of X. The set of minimal elements of a subset X will be denoted by min X.

PROPOSITION 6.1.3. *The following conditions on a partially ordered set E are equivalent*:

(i) *The ideals of E are finitely generated.*
(ii) *The ascending chain condition holds for the ideals of E.*
(iii) *Every infinite sequence of elements of E has an infinite ascending sub-sequence.*
(iv) *Every infinite sequence of elements of E has an ascending sub-sequence of length 2.*

(v) *There exist in E neither an infinite strictly descending sequence nor an infinite set of pairwise incomparable elements.*

(vi) *For every nonempty subset X of E, min X is nonempty and finite.*

Proof. The implication (i) $\Rightarrow$ (ii) is a classical phenomenon: if $\{I_n\}$, for n in $\mathbb{N}$, is an ascending chain of ideals, their union I is an ideal, finitely generated by hypothesis, and its generators belong all to a certain I_m; the chain is stationary from m on. (ii) $\Rightarrow$ (iii): Let $S = s_1, s_2, \dots, s_n, \dots$ be an infinite sequence; (ii) implies that there exists an integer k such that S is included in the ideal generated by $\{s_1, s_2, \dots, s_k\}$. There must then be at least one value $j, 1 \leqslant j \leqslant k$, such that there exist an infinity of s_i greater than s_j. The procedure begins again with those s_i, and s_j is the first element of an infinite ascending sub-sequence of S. (iii) $\Rightarrow$ (iv), and (iv) $\Rightarrow$ (v) are obvious. (v) $\Rightarrow$ (vi): Let X be a nonempty subset of E; for any x in X there must be a minimal element of X smaller than x since there are no infinite descending sequences, and thus min X is nonempty; it is finite since its elements are pairwise incomparable. (vi) $\Rightarrow$ (i) since for any ideal I, min I is a set of generators of I. ∎

Proposition 6.1.3 is due to Higman (1952); note that we proved Theorem 6.1.2 by establishing property (v) for the division ordering (see Problem 6.1.2 for other applications of Proposition 6.1.3).

Let us now come back to A^*, ordered by division. There is an alternative way for defining division. Although it does not bring anything new, it offers a new point of view that may often be convenient to have in mind.

The *shuffle* of two words f and g of A^* is the subset of A^*, denoted by $f \circ g$ and defined by:

$$f \circ g = \{h \mid h = f_1 g_1 f_2 g_2 \cdots f_n g_n, n \geqslant 0,$$

$$f_i, g_i \in A^*, f = f_1 f_2 \cdots f_n, g = g_1 g_2 \cdots g_n\}$$

The *shuffle of two subsets* X and Y of A^* is the union of the shuffles of each element of X with each element of Y and is denoted by $X \circ Y$. One verifies that the shuffle is a commutative and associative operation on $\mathscr{P}(A^*)$; see Problem 6.1.12 for an alternative definition of the shuffle.

It is clear that a word g divides a word f if, and only if, there exists a word h such that f belongs to $g \circ h$; that is

$$\forall f, g \in A^* \quad g \mid h \Leftrightarrow f \in g \circ A^* \tag{6.1.1}$$

A subset X of A^* will be called a *shuffle ideal* if $X = X \circ A^*$; since the shuffle is associative, the *shuffle ideal generated* by a subset X is $X \circ A^*$. A shuffle ideal is *principal* if it is generated by one word. From (6.1.1) it follows that the shuffle ideals are exactly ideals of the ordered set A^* and that $\overline{X} = X \circ A^*$. In order to avoid confusion between the ideals of the monoid A^*, and the ideals of the ordered set A^*, from now on we call the latter *shuffle ideals*.

As a consequence of Theorem 6.1.2, and Proposition 6.1.3, *any shuffle ideal is finitely generated*. For future reference we state this in the following equivalent form:

COROLLARY 6.1.4. *The boolean algebra generated by the principal shuffle ideals is equal to the boolean algebra generated by the shuffle ideals.*

Example 6.1.5. Let X be the set of *square words*; that is, $X = \{ f \mid f = uu, |u| > 1 \}$. Then

$$\overline{X} = \bigcup_{a \in A} (aa \circ A^*)$$

the set of words that contain at least one letter repeated twice.

6.2. Comparing the Subwords

For every integer m, let J_m be the equivalence relation on A^* defined as follows: Two words f and g are equivalent modulo J_m if they have the same set of subwords of length less than, or equal to m. That is

$$\forall f, g \in A^* \qquad f \equiv g [J_m] \Leftrightarrow (\forall s \in A^* |s| \leq m \Rightarrow (s|f \Leftrightarrow s|g)). \quad (6.2.1)$$

In this section, we shall study the structure of the classes of the equivalence J_m.

Example 6.2.1. J_0 is the universal relation on A^*; J_1 coincides with the mapping equivalence of the function alph:

$$f \equiv g [J_1] \Leftrightarrow \mathrm{alph}(f) = \mathrm{alph}(g).$$

Let $S(m, f)$ denote the set of subwords of length less than or equal to m of a word f, and let $NS(m, f)$ denote its complement within

$$A^{\leq m} = \bigcup_0^m A^n.$$

Thus (6.2.1) may be written as

$$f \equiv g[J_m] \Leftrightarrow S(m, f) = S(m, g),$$

which shows clearly that J_m is indeed an equivalence relation. The class of f modulo J_m will be denoted by $[f]_m$.

Remark 6.2.2. Let m be a fixed integer. By the length argument, if $|f| < m$ then $[f]_m$ is a singleton (it may not be a singleton if $|f| = m$; see Problem 6.2.1). Suppose now that $|f| \geqslant m$ and let g be a word that has the same set of subwords of *length* m as f, then f and g have the same set of subwords of *length less than or equal to* m and g is equivalent to f modulo J_m. Hence (6.2.1) splits into two relations:

$$\forall f, g \in A^* \quad |f| < m \Rightarrow (f \equiv g[J_m] \Leftrightarrow g = f), \qquad (6.2.2)$$

$$\forall f, g \in A^* \quad f \equiv g[J_m] \Leftrightarrow (\forall s \in A^m \, s | f \Leftrightarrow s | g). \qquad (6.2.3)$$

We have the following properties of the relations J_m:

PROPOSITION 6.2.3. *For n smaller than m, J_n is coarser than J_m.*   ∎

PROPOSITION 6.2.4. *For every m, J_m is a congruence on A^* of finite index.*

Proof. Let f, g, u, and v be words in A^*. Any subword w of ufv of length m factors into $w = w_1 w_2 w_3$ with $w_1 | u, w_2 | f$, and $w_3 | v$ and then $|w_2| \leqslant m$; hence w is a subword of ugv if $f \equiv g[J_m]$ and J_m is a congruence. Moreover the index of J_m is equal to the number of distinct subsets $S(m, f)$ when f ranges over A^* and hence is clearly bounded by 2^l where

$$l = \frac{\operatorname{Card}(A)^{m+1} - 1}{\operatorname{Card}(A) - 1} = \operatorname{Card}(A^{\leqslant m}). \qquad ∎$$

Observe that not any subset of $A^{\leqslant m}$ is apt to be an $S(m, f)$ for some f in A^*; for example, there exists no f in $\{a, b\}^*$ such that $S(2, f) = \{a, bb\}$ since if a and bb are contained in $S(2, f)$ so is b and at least one of the two words ab and ba. Hence 2^l is a strict bound for the index of J_m; actually neither a suitable characterization of the sets $S(m, f)$ nor an explicit function of m and $\operatorname{Card}(A)$ giving the index of J_m are known.

Now, the boolean closure of shuffle ideals is characterized in terms of the classes modulo J_m.

PROPOSITION 6.2.5. *The family of subsets of A^* that are union of classes modulo J_m for some integer m is equal to the boolean closure of shuffle ideals.*

Proof. Let X be a union of classes modulo J_m for some integer m. Since J_m is of finite index, the set X is a finite union of such classes. And the class modulo J_m of a word $f \in A^*$ may be written

$$[f]_m = (\bigcap_{s \in S(m,f)} s \circ A^*) \backslash (\bigcup_{t \in NS(m,f)} t \circ A^*)$$

which belongs to the boolean closure of shuffle ideals.

Conversely, for each f in A^*, the shuffle ideal generated by f is saturated modulo J_m for each $m \geqslant |f|$. By Corollary 6.1.4, this proves that any element of the boolean closure of shuffle ideals is saturated modulo J_m, for some m: indeed, it is enough to choose m greater than the maximum of the lengths of the generators of the principal ideals involved in the construction of X. ∎

The main result on the classes of the congruences J_m is given by the following theorem, which appeared first in Simon 1975.

THEOREM 6.2.6. *Let m be an integer and let f and g be two words of A^*, equivalent modulo J_m. Then there exists a word h such that both f and g divide h and such that h is equivalent to f, and to g, modulo J_m.*

In other words, any two elements in an equivalence class of J_m have a common upper bound (relative to division) in that equivalence class. Or, in order to stress the unexpected fact that is stated by Theorem 6.2.6: For any two words f and g that have the same set of subwords of length m, it is possible to find a word h that contains both f and g as subwords and that has no more subwords of length m than f (and g); note that the word h is, in general, longer than either f or g.

There are cases where the theorem obviously holds: when f divides g, or vice versa, or when f and g contain as subwords all words of length m; in the latter case any element h of the shuffle of f and g for instance would do. But of course the theorem remains to be proved in the general case.

Example 6.2.7. $m = 4$ $\quad f = a^2b^4a^4b^2 \quad g = a^3b^3a^3b^3 \quad h = a^3b^4a^4b^3$.

Our proof of Theorem 6.2.6 follows the main line of the one in Simon 1975 but the use of a new notion, the *subword distance*, will make it more concise. We thus postpone it until after the definition of the subword distance. But we may derive immediately some consequences of Theorem 6.2.6:

COROLLARY 6.2.8. *For each integer m, any equivalence class modulo J_m is either a singleton or infinite.*

Proof. Suppose that a word f is not isolated in its class modulo J_m. From Theorem 6.2.6 it follows that there exists a word h such that $f \neq h$, $f \mid h$, and $f \equiv h[J_m]$. Thus there exist u, and v in A^*, a in A such that $h = uav$ and $f \mid uv$. We therefore have

$$uv \equiv uav[J_m]. \tag{6.2.4}$$

Now we claim that $uv = ua^n v[J_m]$ for every integer $n \geqslant 0$; it is enough indeed to show that $uav \equiv uaav[J_m]$ and the claim will follow by induction.

Assume that there exists a word w of length m such that $w \mid uaav$ and $w \nmid uav$. Therefore w must factor into $w = saat$ with $s \mid u$ and $t \mid v$. The word sat then divides uav and is of length $m - 1$; by (6.2.4) it divides uv. But then either $sa \mid u$, or $at \mid v$ and in both cases $saat \mid uav$: a contradiction which completes the proof. ∎

We observe also that, contrary to what happens for upper bounds, two words f and g in a congruence class of J_m might not have a common lower bound (relative to division) in that congruence class. Indeed, $ab \equiv ba[J_1]$ but both ab and ba are minimal words in $[ab]_1$. Actually the set of minimal elements of a congruence class J_m was characterized in Simon 1972, where the following theorem is proved:

THEOREM 6.2.9. *Given a word f in A^* and an integer $m \geqslant 0$, there exist an integer $k \geqslant 0$, words $u_0, u_1, \ldots, u_k$ in A^*, and subsets $B_1, B_2, \ldots, B_k$ of A such that the set $\min[f]_m$ is given by*

$$\min[f]_m = \{ u_0 v_1 u_1 v_2 u_2 \cdots v_k u_k \mid v_i \text{ is a permutation of } B_i \}.$$

Here we say that v in A^* is a permutation of a subset B of A if $\text{alph}(v) = B$ and $|v| = \text{Card}(B)$.

The definition of the congruences J_m allows definition of a distance on A^* by means of the following: First let $\hat{N} = N \cup \{\infty\}$ be the completion of N as an upper-semilattice. The element ∞ is the (least) upper bound of any unbounded subset of N.

For all f and g in A^*, let $\delta(f, g)$ be the element of $\hat{N}$ defined by

$$\delta(f, g) = \max\{ m \in N \mid f \equiv g[J_m] \}.$$

Clearly $\delta(f, g) = \infty$ if and only if $f = g$, and since for each m and n, J_m is an equivalence relation coarser than J_{m+n}, it follows that:

$$\forall f, g, h \in A^* \qquad \delta(f, g) \geqslant \min\{\delta(f, h), \delta(g, h)\}. \tag{6.2.5}$$

From these two remarks one deduces that the function θ defined by $\theta(f, g) = 2^{-\delta(f, g)}$ is an *ultrametric distance* on A^* (since relation (6.2.5) implies that $\theta(f, g) \leqslant \max\{\theta(f, h), \theta(g, h)\}$). Actually we shall be concerned in the sequel only with the function δ, and by abuse of language we call δ itself the *subword distance*, or even, for short, the *distance*, between f and g. The triangular relation (6.2.5) for the distance δ takes the following form (classical for ultrametric distances, with *largest* instead of *smallest*):

PROPOSITION 6.2.10. *Let f, g, and h be in A^*. Of the three numbers $\delta(f, g), \delta(g, h)$, and $\delta(f, h)$, the two smallest ones are equal.*

This may be written as:

$$\forall f, g, h \in A^* \qquad \delta(f, g) \leqslant \delta(f, h) \Rightarrow \delta(f, g) = \min\{\delta(f, h), \delta(g, h)\},$$
$$(6.2.6)$$

or as

$$\delta(f, g) \geqslant m \quad \text{and} \quad \delta(g, h) \geqslant m \Rightarrow \delta(f, h) \geqslant m,$$

or it may be rephrased as (with respect to the "distance" δ): Every triangle has two equal "short" sides.

Proof. First rewrite the relation (6.2.5) with f exchanged with h:

$$\delta(g, h) \geqslant \min\{\delta(f, h), \delta(f, g)\}.$$
$$(6.2.7)$$

Now the hypothesis

$$\delta(f, g) \leqslant \delta(f, h)$$
$$(6.2.8)$$

matched with (6.2.7) gives

$$\delta(f, g) \leqslant \delta(g, h)$$
$$(6.2.9)$$

which in turn, by (6.2.5), gives

$$\delta(f, g) = \min\{\delta(f, h), \delta(g, h)\}. \qquad \blacksquare$$

We can now state Theorem 6.2.6 under the following form

THEOREM 6.2.11. *For any two words f and g in A^* there exists a word h in A^*, such that both f and g divide h and that $\delta(f, g) = \min\{\delta(f, h), \delta(g, h)\}$.*

The two statements are indeed equivalent: Theorem 6.2.11 implies Theorem 6.2.6, for if $f \equiv g[J_m]$ then $\delta(f, g)$ is greater than or equal to m and by Theorem 6.2.11 so are both $\delta(f, h)$ and $\delta(g, h)$; that is, $f \equiv h \equiv g[J_m]$. Theorem 6.2.6 implies Theorem 6.2.11 since if we take $m = \delta(f, g)$ Theorem

6.2.6 implies that $f \equiv h \equiv g[J_m]$; that is, both $\delta(f, h)$ and $\delta(g, h)$ are greater than or equal to $\delta(f, g)$.

In order to prove Theorem 6.2.11 we shall establish a sequence of three lemmas and, before that, we present some properties of the division, the easy verification of which is left to the reader. They will be used implicitly in the proofs to come (and have been used in the proof of Corollary 6.2.8).

Let u, v, s, and t be words in A^*, and a and b letters in A. Then:

$$st \mid uv \quad \Rightarrow \quad s \mid u \quad \text{or} \quad t \mid v,$$
$$sat \mid uv \quad \Rightarrow \quad sa \mid u \quad \text{or} \quad at \mid v,$$
$$at \mid bv \quad \text{and} \quad a \neq b \quad \Rightarrow \quad at \mid v.$$

We shall also say that a word z in A^* *distinguishes* words f and g of A^* if z divides exactly one of f or g. It is clear that for distinct words f and g, any shortest word distinguishing f and g has length $\delta(f, g) + 1$.

LEMMA 6.2.12. *Let u be a word in A^* and let a be a letter in A. Then every shortest word z that distinguishes ua and u has a factorization $z = sa$, with $|s| = \delta(ua, u)$.* ∎

LEMMA 6.2.13. *Let u and v be words in A^* and let a be a letter in A. Then*

$$\delta(uav, uv) = \delta(ua, u) + \delta(av, v). \tag{6.2.10}$$

Proof. Let sa (resp. at) be a shortest word that distinguishes ua and u (resp. av and v) as given by Lemma 6.2.12 (resp. its dual); the word sat distinguishes uav and uv and thus

$$\delta(uav, uv) \leq \delta(ua, u) + \delta(av, v).$$

On the other hand, a word f distinguishing uav and uv must factorize into $f = sat$ where sa distinguishes ua and u and where at distinguishes av and v. This implies

$$\delta(ua, u) + \delta(av, v) \leq \delta(uav, uv) \qquad \blacksquare$$

The key idea of the proof of Theorem 6.2.11 lies in the third lemma.

LEMMA 6.2.14. *Let u, v, and w be words in A^* and let a and b be distinct letters in A. Then*

$$\delta(uav, ubw) \leq \max\{\delta(ubav, uav), \delta(uabw, ubw)\}. \tag{6.2.11}$$

Proof. Adjust notations, changing v and w, as well as a and b, if necessary, so that

$$\delta(abw, bw) \le \delta(bav, av).\tag{6.2.12}$$

By Lemma 6.2.12 and its dual, there exist r, s, and t in A^* such that

$\quad |r| = \delta(ua, u)$ and ra distinguishes ua and u,

$\quad |s| = \delta(ub, u)$ and sb distinguishes ub and u,

$\quad |t| = \delta(abw, bw)$ and at distinguishes abw and bw.

If t divides v, then clearly rat divides uav but does not divide ubw. Thus

$$\delta(uav, ubw) \le |rat| - 1 = \delta(ua, u) + \delta(abw, bw).$$

By Lemma 6.2.13

$$\delta(ua, u) + \delta(abw, bw) = \delta(uabw, ubw);$$

hence the lemma holds in this case.

Assume now that t does not divide v. To begin with, we claim that

$$\delta(uav, ubw) \le |s| + |t|.\tag{6.2.13}$$

Indeed if $t|w$ then $sbt|ubw$. But $sbt \nmid uav$; that is, sbt distinguishes uav and ubw. Hence the claim holds in this case. Now if $t \nmid w$, then recalling that $t|bw$, it follows that $t = bt'$ for some t' such that $t'|w$. Then $st = sbt'$ divides ubw but does not divide uav since $a \ne b$, $t \nmid v$, and $sb \nmid u$. Then st distinguishes uav and ubw, and hence Eq. (6.2.13) holds again.

Finally (6.2.13) may be written as

$$\delta(uav, ubw) \le \delta(ub, u) + \delta(abw, bw),$$

and then (6.2.12) implies

$$\delta(uav, ubw) \le \delta(ub, u) + \delta(bav, av),$$

which gives, by Lemma 6.2.13

$$\delta(uav, ubw) \le \delta(ubav, uav)$$

and the proof of the lemma is complete.   ∎

Proof of Theorem 6.2.11. We prove the assertion by induction on the integer $d(f, g) = |f| + |g| - 2|f \wedge g|$, where $f \wedge g$ denotes the longest common left factor of f and g. Initially we note that if $f \mid g$ or $g \mid f$, then the assertion holds with $h = g$ or $h = f$, respectively. This gives, in particular, the proof of the basis of the induction. Assuming that $f \nmid g$ and $g \nmid f$, the cases where f is a left factor of g or where g is a left factor of f are excluded, and there exist u, v, and w in A, and a and b in A, such that $a \neq b$, $f = uav$, and $g = ubw$. By Lemma 6.2.14

$$\delta(f, g) \leq \max\{\delta(ubav, f), \delta(uabw, g)\}.$$

Adjust notations, changing v and w, as well as a and b, if necessary, so that

$$\delta(f, g) \leq \delta(ubav, f).$$

Then, by Proposition 6.2.10

$$\delta(f, g) = \min\{\delta(ubav, f), \delta(ubav, g)\}. \tag{6.2.14}$$

On the other hand, $a \neq b$, hence

$$d(ubav, g) = d(ubav, ubw) \leq |av| + |w|,$$

and thus

$$d(ubav, g) < |av| + |bw| = d(f, g).$$

Thus, by the induction hypothesis, there exists h in A^* such that $ubav$ and g both divide h and

$$\delta(ubav, g) = \min\{\delta(ubav, h), \delta(g, h)\}. \tag{6.2.15}$$

Finally, note that f divides $ubav$, and then h. Hence, every word that distinguishes $ubav$ and h also distinguishes f and h. It follows that $\delta(f, h) \leq \delta(ubav, h)$, and again by Proposition 6.2.10:

$$\delta(f, h) = \min\{\delta(ubav, h), \delta(ubav, f)\}. \tag{6.2.16}$$

Now, (6.2.15) in (6.2.14) give

$$\delta(f, g) = \min\{\delta(ubav, f), \delta(ubav, h), \delta(g, h)\}, \tag{6.2.17}$$

and (6.2.16) in (6.2.17) give the expected equality:

$$\delta(f, g) = \min\{\delta(f, h), \delta(g, h)\}.$$

Since by induction g divides h and, as already noted, f also divides h, the proof is complete. ∎

The proof of Theorem 6.2.11 we have given here will be a constructive one as soon as we shall have an effective procedure to compute $\delta(av, v)$ and a shortest word that distinguishes between av and v. Such a procedure is given by the following proposition and its proof.

PROPOSITION 6.2.15. *Let t and v be two words of A^*, with t not equal to the empty word, and let m be a positive integer. Let $P(m, t, v)$ be the following predicate over $\mathbb{N} \times A^+ \times A^*$: there exists a factorization of v into m factors $v = v_1 v_2 \cdots v_m$ such that*

$$\text{alph}(t) \subseteq \text{alph}(v_1) \subseteq \cdots \subseteq \text{alph}(v_m) \qquad (6.2.18)$$

and, by convention, $P(0, t, v)$ holds for any couple (t, v) of $A^+ \times A^*$. We then have $\delta(tv, v) = max\{m \mid P(m, t, v)\}$.

Proof. We show first the implication

$$P(m, t, v) \Rightarrow (\forall s \in A^m \quad s \mid tv \Rightarrow s \mid v). \qquad (6.2.19)$$

Let $v = v_1 v_2 \cdots v_m$ be a factorization of v that satisfies (6.2.18) and let $s = s_1 s_2 \cdots s_m$, s_i in A, be a subword of tv of length m. If s does not divide v there exists a greatest integer j, between 1 and m, such that

$$s_j s_{j+1} \cdots s_m \nmid v_j v_{j+1} \cdots v_m.$$

By (6.2.18) alpha$(tv) \subseteq$ alph(v_m), hence $s_m \mid v_m$. Thus, $j < m$ and the choice of j implies that

$$s_{j+1} \cdots s_m \mid v_{j+1} \cdots v_m.$$

These two assertions imply that s_j does not belong to alph(v_j). On the other hand, since by hypothesis $s_j s_{j+1} \cdots s_m$ divides tv, they also imply that s_j belongs to alph$(tv_1 v_2 \cdots v_{j-1})$, which is contained, by (6.2.18), in alph(v_j), a contradiction. Hence s divides v.

To each pair (t, v) of $A^+ \times A^*$ we now associate an integer $k(t, v)$ and a word $h(t, v)$ by the following definition:

- If alph$(t) \not\subseteq$ alph(v), then $k(t, v) = 0$ and $h(t, v)$ is any letter in alph$(t) \setminus$ alph(v);
- If alph$(t) \subseteq$ alph(v), then let t' be the shortest left factor of v such that alph$(t) \subseteq$ alph(t') and let v' be the right factor of v such that $v = t'v'$. Since t is not the empty word, neither is t' and $t' = t_1 a$ with a in A. We define then $k(t, v) = k(t', v') + 1$ and $h(t, v) = ah(t', v')$.
  We claim now that

$$h(t, v) \mid tv \qquad \text{and} \qquad h(t, v) \nmid v, \qquad (6.2.20)$$

which will be proved by induction on $k(t, v)$ and which trivially holds for $k(t, v) = 0$.

We note first that the minimality of the factor $t' = t_1 a$ implies that the letter a belongs to alph(t) but does not belong to alph(t_1). Hence if $h(t', v')$ divides $t'v'$, then $h(t, v) = ah'(t', v')$ divides $tt'v' = tv$. On the other hand if $ah(t', v')$ divides $v = t_1 av'$, then $h(t', v')$ divides v' also; the claim is established.

It is routine to verify, again by induction on $k(t, v)$, that

$$P(k(t, v), t, v) \tag{6.2.21}$$

holds and that

$$|h(t, v)| = k(t, v) + 1. \tag{6.2.22}$$

This completes the proof; indeed,

- (6.2.21) implies that $k(t, v) \leqslant \max\{m \mid P(m, t, v)\}$;
- (6.2.19) implies that $\max\{m \mid P(m, t, v)\} \leqslant \delta(tv, v)$ and (6.2.20) and (6.2.22) together imply that $\delta(tv, v) < k(t, v) + 1$. ∎

From Proposition 6.2.15 follows immediately:

COROLLARY 6.2.16. *For any word f and g in A^* and for any integer m we have $(fg)^m \equiv g(fg)^m [J_m]$.*

Theorem 6.2.6 and Proposition 6.2.15 (via Corollary 6.2.16) are the key arguments for the characterization of the quotients of A^* by the congruences J_m: cf. Simon 1975 and Problem 6.2.9. This characterization was first given in Simon 1972 with a more complicated proof. The presentation of it in Eilenberg 1976 is modeled after the latter reference.

We now turn to the problem of characterizing a word f among the other words of same length by the set of its subwords of a fixed length m. The solution is given by the following results, due to Schützenberger and Simon and that appear here for the first time.

THEOREM 6.2.16. *Let A be an alphabet with at least two letters, and let m and n be two integers. The restriction of the equivalence J_m to A^n is the identity if, and only if, the inequality $n \leqslant 2m - 1$ holds.*

The example of the two words $a^{m-1}ba^m$ and $a^m ba^{m-1}$ that have the same set of subwords of length m shows that the condition is necessary. That it is sufficient is a consequence of the two following lemmas (recall that $S(m, f)$ denotes the set of subwords of f of length less than, or equal to m). Note that the lemmas give a reconstruction of a word f from a (strict) subset of $S(m, f)$.

LEMMA 6.2.17. *Let $A = \{a, b\}$ be a two-letter alphabet. Let m be an integer and let $n = 2m - 1$. Every word f of A^* of length less than, or equal to n is uniquely characterized by its length and by the set*

$$D(f) = S(m, f) \cap (a^*b^* \cup b^*a^*).$$

Proof. For a word f, we define

$$D(f) = S(m, f) \cap (a^*b^* + b^*a^*).$$

Let us assume, for a contradiction, that f and g are different words in A^*, such that

$$|f| = |g| \leq n \quad \text{and} \quad D(f) = D(g) = D.$$

We further define

$$p = \max\{|s| \,|\, s \in D \cap a^*\}$$

and

$$q = \max\{|s| \,|\, s \in D \cap b^*\}.$$

Adjust notation, changing a and b if necessary, so that $q \leq p$. Then clearly, $|f|_a \geq p$ and $|f|_b \geq q$; hence

$$2q \leq p + q \leq |f|_a + |f|_b = |f| \leq n = 2m - 1.$$

Since q is an integer it follows that $q < m$, and this implies that $|f|_b = q$. Similarly, $|g|_b = q$. Thus, there exist integers $i_0, i_1, \ldots, i_q, j_0, j_1, \ldots, j_q$, such that

$$f = a^{i_0} b a^{i_1} b \cdots a^{i_{q-1}} b a^{i_q}$$

and

$$g = a^{j_0} b a^{j_1} b \cdots a^{j_{q-1}} b a^{j_q}.$$

Since $f \neq g$ by assumption, there exists a smallest k, such that $i_k \neq j_k$. Adjust notation, changing f and g if necessary, so that $i_k < j_k$. Since $|f| = |g|, i_l = j_l$ for $0 \leq l < k$ and $i_k < j_k$ it follows that

$$j_{k+1} + \cdots + j_q < i_{k+1} + \cdots + i_q.$$

Let s and t be given by

$$s = a^{i_0 + i_1 + \cdots + i_k + 1} b^{q-k} \quad \text{and} \quad t = b^{k+1} a^{j_{k+1} + \cdots + j_q + 1}.$$

Clearly $s \mid g$, $s \nmid f$, $t \mid f$ and $t \nmid g$. Hence since both s and t belong to $a^*b^* + b^*a^*$, it follows that $|s|, |t| > m$. Then

$$i_0 + i_1 + \cdots + i_k + q - k \geqslant m$$

and

$$k + 1 + j_{k+1} + \cdots + j_q \geqslant m.$$

Summing these two inequalities and recalling that $i_l = j_l$, for $0 < l < k$, and that $i_k < j_k$, we have:

$$j_0 + j_1 + \cdots + j_q + q = |g| \geqslant 2m,$$

a contradiction. Thus, $i_k = j_k$ for every k and $f = g$. This concludes the proof. ∎

Remark 6.2.18. Two distinct words f and g of $\{a, b\}^*$ of length less than or equal to $2m - 1$ may be equivalent modulo J_m, e.g. $m = 4$ and $abaaba \equiv abaaaba[J_4]$. But it is noteworthy that under these hypothesis (i.e., $f \equiv g[J_m]$, $|f| \leqslant 2m - 1, |g| \leqslant 2m - 1$) the beginning of the proof of Lemma 6.2.17 implies that if $|f|_a \geqslant |f|_b$ then $|g|_a \geqslant |g|_b$ and $|f|_b = |g|_b$, and this last value is determined by the set D, which is common to f and g, even when the lengths f and g are not known.

LEMMA 6.2.19. *Let A be any alphabet and let f be a word of A^*. For every two element subset $\{a, b\}$ of A let $f_{a,b}$ be the longest subword of f that belongs to $\{a, b\}^*$. The word f is then uniquely characterized by the set $L = \{f_{a,b} \mid a, b \in A\}$.*

Proof. The first letter x of f is characterized by the condition that, for all a in A, $f_{a, x}$ begins with x; clearly there is exactly one letter x satisfying this condition. More generally, if h the left factor of length k of f is known, the $(k + 1)$th letter x of f is characterized by the condition that for all a in A, x is the $(|h|_a + |h|_x + 1)$th letter of $f_{a, x}$. ∎

Proof of Theorem 6.2.16. Let m be an integer and let f be a word of A^* of length less than or equal to $2m - 1$. By Lemma 6.2.19 f is determined by the set of subwords $f_{a, b}$. Each $f_{a,b}$ is, a fortiori, of length less than or equal to $2m - 1$ and then it is characterized by the set $S(m, f) \cap \{a, b\}^*$ as soon as its length is known. From the preceding Remark follows that the smaller of the integers $|f_{a,b}|_a = |f|_a$ and $|f_{a,b}|_b = |f|_b$ is determined by the set $S(m, f)$. By inspection of all the pairs of letters, $S(m, f)$ determines all the numbers $|f|_a$ but the largest one, which can be obtained by difference with $|f|$.   ∎

The proof just given for Theorem 6.2.16 using Lemmas 6.2.17 and 6.2.19 is in fact the demonstration of the following stronger result:

PROPOSITION 6.2.20. *Any word f of A^* is characterized by its length and by the set of its subwords that belong to the set*

$$T = \bigcup_{\substack{a, b \in A^* \\ a \neq b}} a^* b^*$$

and the length of which is less than or equal to

$$m = \left\lceil \frac{n+1}{2} \right\rceil,$$

where n is defined by $n = \max\{|f|_a + |f|_b \mid a, b \in A, a \neq b\}$.

6.3.   Counting the Subwords

In the very beginning of this chapter, we came back to the definition of a word as a finite sequence of letters, and we defined a subword of a word f to be a sub-sequence of f. We also noticed that two sub-sequences of f, distinct as sub-sequences, may define the same word. For instance in the word *bacbcab* there is only one sub-sequence equal to *aba*, but there are two of them equal to *aca*, three to *ab*, and four to *bab*. Until now we have been interested only in the fact whether a word g appears as a sub-sequence in a word f or not. In this section we shall consider also the number of times g is a subword of f. Given words f and g in A^*, the number of distinct sub-sequences of f that are equal to g is called the *binomial coefficient* of f and g and is denoted by $\binom{f}{g}$.

Example 6.3.1. $\binom{abab}{ab} = 3,$ $\binom{aabbaa}{aba} = 8.$

Let f and g be two words written with only one letter a and of lengths p and q respectively; then $\binom{a^p}{a^q} = \binom{p}{q}$ where $\binom{p}{q}$, with integers p and q, denotes the classical binomial coefficient. This remark justifies our terminology.

The empty word is defined to be the sequence of length 0 and there is exactly one sub-sequence of length 0 in any sequence. Then

$$\forall f \in A^* \qquad \binom{f}{1} = 1. \qquad (6.3.1)$$

By the length argument we have

$$\forall f, g \in A^* \qquad |f| < |g| \Rightarrow \binom{f}{g} = 0. \qquad (6.3.2)$$

Recall that, for all f and g in A^*, the *Kronecker symbol* $\delta_{f,g}$ has value 1 if $f = g$, and 0 otherwise. We then have

PROPOSITION 6.3.2. *The following holds*

$$\forall f, g \in A^*, \forall a, b \in A, \qquad \binom{fa}{gb} = \binom{f}{gb} + \delta_{a,b} \binom{f}{g}. \qquad (6.3.3)$$

Proof. If $a \neq b$, then every sub-sequence of fa that is equal to gb is a sub-sequence of f, and conversely any such sub-sequence would do. The formula then holds for that case. If $a = b$, the sub-sequences of fa that are equal to ga fall in two classes: those that do not contain the last element of the sequence fa and those that do. The first ones, as before, are sub-sequences of f equal to ga, and conversely any such sub-sequence would do. The latter ones, after cutting their last element, become sub-sequences of f, equal to g, and conversely any such sub-sequence of f, when added the last element of the sequence fa, is equal to ga. By construction no sub-sequence may belong to both classes and the formula is proved. ∎

Proposition 6.3.2 is the generalization of Pascal's formula

$$\binom{p}{q} = \binom{p-1}{q} + \binom{p-1}{q-1}$$

to our binomial coefficients; we indeed used the same argument to prove it as the one that is commonly used for establishing Pascal's formula. Using suitable notation, it is possible to make (6.3.3) look even more like Pascal's formula (cf. Problem 6.3.3)

PROPOSITION 6.3.3. *Relations (6.3.1)–(6.3.3) determine the binomial coefficients* $\binom{f}{g}$ *for every f and g in A*.*

Proof. By induction on the length of f. Relations (6.3.1) and (6.3.2) determine $\binom{1}{g}$ for all g in A^* (indeed the relation $\binom{1}{g} = 0$ for all g of length greater than 0 would have been sufficient instead of (6.3.2)); and then $\binom{fa}{g}$ is determined for all g in A^*: by (6.3.1) for $g=1$ and by (6.3.3) for every g of length greater than 0.   ∎

Remark 6.3.4. As usual with the classical binomial coefficients, we can order the binomial coefficients in an infinite matrix, the rows and the columns of which are indexed by A^* and such that the entry (f, g) has value $\binom{f}{g}$. If rows and columns are both lexicographically ordered, the matrix is lower triangular; we call it the Pascal matrix P.

Remark 6.3.5. If $\mathbb{K}$ is any homomorphic image of $\mathbb{N}$, with $\varphi: \mathbb{N} \to \mathbb{K}$, then (6.3.1)–(6.3.3) can be used to define $\mathbb{K}$-binomial coefficients $\binom{f}{g}_{\mathbb{K}}$ and then $\varphi\binom{f}{g} = \binom{f}{g}_{\mathbb{K}}$ holds. For instance if $\mathbb{B}$ is the boolean semi-ring $\binom{f}{g}_{\mathbb{B}} = 1$ if, and only if $\binom{f}{g} \neq 0$, that is, if and only if $g \mid f$.

We shall now extend Newton's formula to our binomial coefficients. Recall that $\mathbb{Z}\langle A \rangle$ (resp. $\mathbb{N}\langle A \rangle$) denotes the algebra (resp. semi-algebra) of polynomials in noncommutative indeterminates in A with integer (resp. positive integer) coefficients. We shall call *Magnus transformation* the algebra endomorphism μ of $\mathbb{Z}\langle A \rangle$ defined by

$$\mu(a) = 1 + a \quad \text{for every } a \text{ in } A$$

The restriction of μ to $\mathbb{N}\langle A \rangle$ is a semi-algebra endomorphism of $\mathbb{N}\langle A \rangle$.

PROPOSITION 6.3.6. *For every f in A^* the following holds:*

$$\mu(f) = \sum_{g \in A} \binom{f}{g} g. \tag{6.3.4}$$

Proof. By induction on the length of f. The equality holds for $f=1$. Let us compute $\mu(fa)$:

$$\mu(fa) = \mu(f)(1+a) = \left(\sum_{h \in A^*} \binom{f}{h} h \right)(1+a)$$

$$= \sum_{h \in A^*} \binom{f}{h} h + \sum_{h \in A^*} \binom{f}{h} ha.$$

We split the first sum into three parts: $h=1$; $h=ga$ for g in A^*; and $h=gb$ for g in A^* and b in A, different from a.

$$\mu(fa) = \binom{f}{1}1 + \sum_{\substack{g\in A^* \\ b\in A, b\neq a}} \binom{f}{gb}gb + \sum_{g\in A^*}\binom{f}{ga}ga + \sum_{h\in A^*}\binom{f}{h}ha.$$

Since h is a dummy index in the third sum, it may be written g, and relation (6.3.3) completes the proof. ∎

The polynomial $\mu(f)$ is sometimes called the *Magnus expansion* of f; Proposition 6.3.6 may also be written as

$$\forall f, g\in A^* \qquad \langle\mu(f), g\rangle = \binom{f}{g}. \qquad (6.3.5)$$

COROLLARY 6.3.7. *For all f, g, and h in A^* one has*

$$\binom{fh}{g} = \sum_{\substack{u, v\in A^* \\ uv=g}}\binom{f}{u}\binom{h}{v}. \qquad (6.3.6)$$

Proof. Since μ is an algebra homomorphism $\mu(fh)=\mu(f)\mu(h)$, and Corollary 6.3.7 expresses the coefficient of g in the product of the two polynomials $\mu(f)$ and $\mu(h)$. ∎

COROLLARY 6.3.8. *For all f and g in A^* one has*

$$\sum_{h\in A^*} (-1)^{|f|+|h|}\binom{f}{h}\binom{h}{g} = \delta_{f,g} \qquad (6.3.7)$$

Proof. Let π be the algebra endomorphism of $\mathbb{Z}\langle A\rangle$, defined by

$$\pi(a) = a-1 = -(1-a) \quad \text{for all} \quad a \quad \text{in} \quad A.$$

Thus, for all a in A, $\pi(a) = -\mu(-a)$; since μ is a homomorphism of algebra, and from Proposition 6.3.6 it follows that for all f in A^*

$$\pi(f) = (-1)^{|f|}\sum_{h\in A^*}(-1)^{|h|}\binom{f}{h}h.$$

Now π and μ are inverse endomorphisms of $\mathbb{Z}\langle A \rangle$ since $\pi(\mu(a)) = \mu(\pi(a)) = a$ for all a in A. Thus

$$\mu(\pi(f)) = f = (-1)^{|f|} \sum_{h \in A^*} (-1)^{|h|} \binom{f}{h} \mu(h)$$

$$= (-1)^{|f|} \sum_{h \in A^*} (-1)^{|h|} \binom{f}{h} \left(\sum_{g \in A^*} \binom{h}{g} g \right)$$

by Proposition 6.3.6. Exchanging the summations completes the proof. ∎

Another way to express corollary 6.3.8 is to state that the inverse of the Pascal matrix P, is the $A^* \times A^*$ matrix Q, the entry (f, h) of which is $(-1)^{|f|+|h|} \binom{f}{h}$. Since the inverse of the transpose of a matrix M is the transpose of the inverse of M we get immediately:

COROLLARY 6.3.9. *For all f and g in A^* one has*

$$\sum_{h \in A^*} (-1)^{|g|+|h|} \binom{f}{h} \binom{h}{g} = \delta_{f,g}.$$ ∎

The orthogonal relation given by Corollary 6.3.8, as well as its companion in Corollary 6.3.9 may be expressed as inverse relations, as is usual when dealing with combinatorial identities (see Riordan 1968). Thus if s and t are real-valued functions on A^* we have the inverse relations

$$s(f) = \sum_{g \in A^*} (-1)^{|g|} (f \quad g) t(g), \qquad t(g) = \sum_{f \in A^*} (-1)^{|f|} \binom{g}{f} s(f)$$

and

$$s(f) = (-1)^{|f|} \sum_{g \in A^*} \binom{g}{f} t(g), \qquad t(g) = (-1)^{|g|} \sum_{f \in A^*} \binom{f}{g} s(f)$$

being aware that the latter may rise convergence problems and is likely to be used with functions that have value zero but on a finite number of words.

In Section 6.1 the shuffle of two words was defined and, by additivity, the shuffle of two subsets of A^*; this notion was in a sense dual to the one of subword. Similarly a suitable product on $\mathbb{Z}\langle\langle A \rangle\rangle$, the module of formal power series on A^* with integer coefficients will provide an adjoint to the Magnus transformation.

The *shuffle product* on the module $\mathbb{Z}\langle\langle A\rangle\rangle$ is the binary operation, denoted by $\circ$, and defined inductively by

$$\forall f \in A^*, \qquad f \circ 1 = 1 \circ f = f \tag{6.3.8}$$

$$\forall a, b \in A, \qquad \forall f, g \in A^*, \qquad fa \circ gb = (f \circ gb)a + (fa \circ g)b \tag{6.3.9}$$

$$\forall s, t \in \mathbb{Z}\langle\langle A^*\rangle\rangle, \qquad s \circ t = \sum_{f, g \in A^*} \langle s, f\rangle\langle t, g\rangle f \circ g \tag{6.3.10}$$

For all f and g in A^*, $f \circ g$ is a homogeneous polynomial of degree $|f| + |g|$, from (6.3.8) and (6.3.9), by induction on $|f| + |g|$. Thus there is no problem with the infinite summation in (6.3.10): the family $\{f \circ g\}_{f, g \in A^*}$ is locally finite.

Shuffle of course bears a close relation to the common idea of shuffling objects. For instance one can see that the coefficient of a word h in the shuffle of f and g is the number of ways one can choose a pair of complementary sub-sequences in h, the first being equal to f and the second to g (cf. Problem 6.3.9).

Example 6.3.10.

$$ab \circ ab = 4aabb + 2abab$$

$$ab \circ ba = abab + 2abba + 2baab + baba.$$

Remark 6.3.11. The sub-semi-module $\mathbb{N}\langle\langle A\rangle\rangle$ of $\mathbb{Z}\langle\langle A\rangle\rangle$ is closed under shuffle product. Observe also that by means of the same relations (6.3.8)–(6.3.10) shuffle product could have been defined on $\mathbb{K}\langle\langle A\rangle\rangle$ for any commutative semi-ring $\mathbb{K}$.

PROPOSITION 6.3.12. *The shuffle product is a commutative and associative operation.*

Both properties, commutativity and associativity, are first established for the elements of A^* by induction on the sum of the lengths of the operands; we leave the verification of this to the reader (see also the proof of Proposition 6.3.15). Relation (6.3.10) then extends both properties to $\mathbb{Z}\langle\langle A\rangle\rangle$. ∎

Since relation (6.3.10) also ensures distributivity of the shuffle over addition in $\mathbb{Z}\langle\langle A\rangle\rangle$, the module $\mathbb{Z}\langle\langle A\rangle\rangle$ equipped with the shuffle becomes an associative and commutative algebra. From (6.3.8), 1 is the unit element of that algebra.

We now come to the property we aimed at when defining shuffle. Recall that $\mathbf{A}^*$ denotes the characteristic series of A^*.

PROPOSITION 6.3.13. *For all g in A^* the following holds*

$$g \circ A^* = \sum_{f \in A^*} \binom{f}{g} f. \tag{6.3.11}$$

Equivalently, Proposition 6.3.13 may also be written as

$$\forall f, g \in A^*, \quad \langle g \circ A^*, f \rangle = \binom{f}{g}, \tag{6.3.12}$$

which shows, as announced, that shuffling with A^* is the adjoint operator of the Magnus transformation. One can say as well that the column of index g in the Pascal matrix gives the coefficients of $g \circ A^*$, whereas the row of index f in the same matrix gives the coefficients of $\mu(f)$.

Relation (6.3.12) extends from A^* to $\mathbb{Z}\langle A \rangle$ by linearity:

$$\forall F, G \in \mathbb{Z}\langle A^* \rangle \quad \langle \mu(F), G \rangle = \langle G \circ A^*, F \rangle \tag{6.3.13}$$

Proof of Proposition 6.3.13. We first remark that from (6.3.8) it follows that

$$\forall f \in A^*, \quad \langle 1 \circ A^*, f \rangle = 1 \tag{6.3.14}$$

and that from the fact that $g \circ h$ is a homogeneous polynomial of degree $|g| + |h|$ it follows that

$$\forall f, g \in A^*, \quad |f| < |g| \Rightarrow \langle g \circ A^*, f \rangle = 0. \tag{6.3.15}$$

We now prove an induction formula for both f and g nonempty. From

$$A^* = 1 + \sum_{c \in A} A^* c$$

follows

$$(gb \circ A^*) = gb + \sum_{c \in A} (gb \circ A^* c)$$

$$= gb + \sum_{c \in A} (g \circ A^* c)b + \sum_{c \in A} (gb \circ A^*)c$$

$$= (g \circ A^*)b + \sum_{c \in A} (gb \circ A^*)c$$

Observe that for any s in $\mathbb{Z}\langle\langle A \rangle\rangle$ and any b in A one has

$$\langle (s)b, fa \rangle = 0 \quad \text{if} \quad a \neq b \quad \text{and}$$

$\langle(s)b, fa\rangle = \langle s, f\rangle$ if $a = b$. Thus,

$$\langle gb \circ \mathbf{A}^*, fa\rangle = \langle(g \circ \mathbf{A}^*)b, fa\rangle + \sum_{c \in A^*} \langle(gb \circ \mathbf{A}^*)c, fa\rangle$$

$$\left.\begin{aligned} &= \langle gb \circ A^*, f\rangle \quad \text{if} \quad a \neq b \\ &= \langle gb \circ \mathbf{A}^*, f\rangle + \langle g \circ \mathbf{A}^*, f\rangle \quad \text{if} \quad a = b \end{aligned}\right\} \quad (6.3.16)$$

and by Proposition 6.3.3, (6.3.12) is established.   ∎

In order to get new conbinatorial identities between binomial coefficients the definition of a new product on $\mathbb{Z}\langle\langle A\rangle\rangle$, very similar to the shuffle product, proved to be useful. It has been introduced in Chen, Fox, and Lyndon 1958.

The *infiltration product*, or *infiltration* for short, is the binary operation on the module $\mathbb{Z}\langle\langle A\rangle\rangle$, denoted by $\uparrow$, and defined inductively by:

$$\forall f \in A^*, \qquad f \uparrow 1 = 1 \uparrow f = f, \tag{6.3.17}$$

$$\forall f, g \in A^*, \qquad \forall a, b \in A,$$

$$fa \uparrow gb = (f \uparrow gb)a + (fa \uparrow g)b + \delta_{a,b}(f \uparrow g)a \tag{6.3.18}$$

$$\forall s, t \in \mathbb{Z}\langle\langle A\rangle\rangle, \qquad s \uparrow t = \sum_{f, g \in A^*} \langle s, f\rangle\langle t, g\rangle(f \uparrow g). \tag{6.3.19}$$

Because of the similarity of the definitions of the shuffle and the infiltration, most of the comments and remarks that have been made for the shuffle also hold for the infiltration. First, the family $\{f \uparrow g\}_{f, g \in A^*}$ is locally finite (see Lemma 6.3.16) so that the summation in (6.3.19) is well defined. Observe also that $\mathbb{N}\langle\langle A\rangle\rangle$ is closed under infiltration.

As for shuffle, infiltration may be given a definition, less suitable for formal and inductive proofs, but which gives a more intuitive idea of the result of the infiltration of two words. The coefficient of a word h in $f \uparrow g$ is the number of pairs of sub-sequences of h that meet the two conditions: (i) they are equal respectively to f and g; (ii) their union gives the whole sequence h (cf. Problem 6.3.9).

Example 6.3.14.

$$ab \uparrow ab = ab + 2aab + 2abb + 4aabb + 2abab,$$

$$ab \uparrow ba = aba + bab + abab + 2abba + 2baab + baba.$$

PROPOSITION 6.3.15. *The infiltration product is commutative and associative.*

Proof. As for shuffle, commutativity and associativity of the infiltration are first established for the elements of A^*, by induction. Since associativity will have a crucial role in the next proposition we give here an explicit proof. The equality $(f \uparrow g) \uparrow h = f \uparrow (g \uparrow h)$ is clear from (6.3.17) as soon as one of the three words f, g, or h is equal to 1. Let thus f, g, and h be words in A^* and a, b, and c be letters in A; one has

$$(fa \uparrow gb) \uparrow hc = ((f \uparrow gb)a) \uparrow hc + ((fa \uparrow g)b) \uparrow hc + \delta_{a,b}((f \uparrow g)a) \uparrow hc$$
$$= ((f \uparrow gb) \uparrow hc)a + (((f \uparrow gb)a) \uparrow h)c + \delta_{a,c}((f \uparrow gb) \uparrow h)a$$
$$+ ((fa \uparrow g) \uparrow hc)b + (((fa \uparrow g)b \uparrow h)c + \delta_{b,c}((fa \uparrow g) \uparrow h)b$$
$$+ \delta_{a,b}[((f \uparrow g) \uparrow hc)a + (((f \uparrow g)a) \uparrow h)c + \delta_{a,c}((f \uparrow g) \uparrow h)a].$$

Grouping together the terms of this sum that end with the letter c we obtain

$$(fa \uparrow gb) \uparrow hc = ((f \uparrow gb) \uparrow hc)a + ((fa \uparrow g) \uparrow hc)b + ((fa \uparrow gb) \uparrow h)c$$
$$+ \delta_{a,c}((f \uparrow gb) \uparrow h)a + \delta_{b,c}((fa \uparrow g) \uparrow h)b$$
$$+ \delta_{a,b}((f \uparrow g) \uparrow hc)a$$
$$+ \delta_{a,b}\delta_{b,c}((f \uparrow g) \uparrow h)a$$

and that expression is symmetric in f, g, and h, as well as in a, b, and c, once the induction hypothesis is applied to the infiltration products within the parenthesis: they involve words such that the sum of their length is strictly less than $|fa| + |gb| + |hc|$, and once it has been noted that $\delta_{a,b}a = \delta_{a,b}b$ and that $\delta_{a,b}\delta_{b,c} = \delta_{a,b}\delta_{b,c}\delta_{c,a}$. ∎

Defining relation (6.3.19) ensures the distributivity of infiltration over the addition, so that the module $\mathbb{Z}\langle\langle A \rangle\rangle$ equipped with the infiltration becomes an associative, and commutative, algebra.

Recall that the *valuation* of an element s of $\mathbb{Z}\langle\langle A \rangle\rangle$, denoted by val($s$), is the smallest integer n such that s contains a monomial of degree n with a nonzero coefficient; by convention val(0) = $+\infty$. We note also deg(P) the degree of an element P of $\mathbb{Z}\langle A \rangle$. A first description of the infiltration of two words is given by the following lemma:

LEMMA 6.3.16. *Let f and g be two words of A^*. The infiltration of f and g is a polynomial $P(f, g)$ of degree $|f| + |g|$. Moreover $f \uparrow g$ may be written as*

$$f \uparrow g = P'(f, g) + f \circ g \ \text{ with } deg(P'(f, g)) < |f| + |g| \quad (6.3.20)$$

If $|f| \geqslant |g|$, $f \uparrow g$ may also be written as

$$f \uparrow g = \binom{f}{g}f + P''(f, g) \ \text{ with } |f| < val(P''(f, g)) \quad (6.3.21)$$

Proof. The first two assertions are easy to prove by induction and are left to the reader. The proof of (6.3.21) is also easy. It goes by double induction on the length of g and then on the length of f. Clearly (6.3.21) holds for $g = 1$ and for all f in A^*. Let us develop

$$fa \uparrow gb = (f \uparrow gb)a + (fa \uparrow g)b + \delta_{a,b}(f \uparrow g)a.$$

Case 1. $|f| \geqslant |gb|$. Then, by induction hypothesis

$$fa \uparrow gb = \left[\binom{f}{gb} f + P''(f, gb) \right] a + \left[\binom{fa}{g} fa + P''(fa, g) \right] b$$

$$+ \delta_{a,b} \left[\binom{f}{g} f + P''(f, g) \right] a$$

$$= \left[\binom{f}{gb} + \delta_{a,b} \binom{f}{g} \right] fa + Q$$

with $\text{val}\, Q > |f| + 1$ and by (6.3.3), relation (6.3.21) holds.

Case 2. $|f| < |gb|$. Then $|f| = |g|$ and it follows:

$$fa \uparrow gb = \left[\binom{gb}{f} gb + P''(gb, f) \right] a + \left[\binom{fa}{g} fa + P''(fa, g) \right] b$$

$$+ \delta_{a,b} \left[\binom{f}{g} f + P''(f, g) \right] a$$

If $f = g$ and $a = b$ then $\delta_{a,b} \binom{f}{g} = 1 = \binom{fa}{gb}$ and (6.3.21) holds; if one of the two preceding conditions is not met then $\delta_{a,b} \binom{f}{g} = 0 = \binom{fa}{gb}$ and (6.3.21) holds again. ∎

COROLLARY 6.3.17. $\forall f, g \in A^*$, $\langle f \uparrow g, f \rangle = \binom{f}{g}.$ (6.3.22)

Relations (6.3.20)–(6.3.22) do not exhaust the relations among binomial coefficients, shuffle, and infiltration. In order to state the one we aim at, we recall one more definition: the *Hadamard product* of two series s and t of $\mathbb{Z}\langle\langle A \rangle\rangle$, denoted by $s \odot t$, is defined by

$$\forall f \in A^*, \qquad \langle s \odot t, f \rangle = \langle s, f \rangle \langle t, f \rangle. \qquad (6.3.23)$$

The Hadamard product may be viewed as a generalization to series (that is, sets with multiplicity) of the operation of intersection on sets; for instance the Hadamard product of two characteristic series s and t is the characteristic series of the intersection of the supports of s and t.

THEOREM 6.3.18. *For all f and g in A^* the following holds*:

$$(f \circ \mathbf{A}^*) \odot (g \circ \mathbf{A}^*) = (f \uparrow g) \circ \mathbf{A}^*. \qquad (6.3.24)$$

Theorem 6.3.18 expresses the equality of two series; if we express instead the equality of their coefficients we obtain

$$\forall f, g, h \in A^*, \qquad \langle f \circ \mathbf{A}^*, h \rangle \langle g \circ \mathbf{A}^*, h \rangle = \langle (f \uparrow g) \circ \mathbf{A}^*, h \rangle,$$

and then, by (6.3.13)

$$\forall f, g, h \in A^*, \qquad \langle f, \mu(h) \rangle \langle g, \mu(h) \rangle = \langle f \uparrow g, \mu(h) \rangle,$$

which may also be written as

$$\forall f, g, h \in A^*, \qquad \binom{h}{f}\binom{h}{g} = \sum_{w \in A^*} \langle f \uparrow g, w \rangle \binom{h}{w}. \qquad (6.3.25)$$

This relation is due to Chen, Fox, Lyndon (1958), who present it in a wider setting. It may be established by induction on the length of the words, as can all other relations of this section, but it is also an immediate consequence of the associativity of the infiltration.

Proof of Theorem 6.3.18. From (6.3.19) we have

$$\forall f, g, h, u \in A^*, \qquad \langle (f \uparrow g) \uparrow h, u \rangle = \sum_{w \in A^*} \langle f \uparrow g, w \rangle \langle w \uparrow h, u \rangle \qquad (6.3.26)$$

and, similarly

$$\forall f, g, h, u \in A^*, \qquad \langle f \uparrow (g \uparrow h), u \rangle = \sum_{v \in A^*} \langle g \uparrow h, v \rangle \langle h \uparrow v, u \rangle. \qquad (6.3.27)$$

Setting $h = u$ in (6.3.26) and using (6.3.22) gives

$$\forall f, g, h \in A^*, \qquad \langle (f \uparrow g) \uparrow h, h \rangle = \sum_{w \in A^*} \langle f \uparrow g, w \rangle \binom{h}{w}$$

We also set $h = u$ in (6.3.27) and then note that Lemma 6.3.16 implies that for all f, g, h, and v in A^* the product $\langle g \uparrow h, v \rangle \langle f \uparrow v, h \rangle$ is equal to zero unless $h = v$. This gives, using again (6.3.22)

$$\forall f, g, h \in A^*, \qquad \langle f \uparrow (g \uparrow h), h \rangle = \binom{h}{f}\binom{h}{g}. \qquad \blacksquare$$

Example 6.3.19. With f and g taken to be letters in A, (6.3.25) becomes

$$\forall h \in A^*, \qquad \forall a, b \in A^*, \qquad \binom{h}{a}\binom{h}{b} = \binom{h}{ab} + \binom{h}{ba}. \quad (6.3.28)$$

By means of the inversion formula (6.3.7) Theorem 6.3.18 gives an explicit formulation of the coefficients of an infiltration product by means of binomial coefficients:

COROLLARY 6.3.20. *For all f, g, and k in A^* one has*

$$\langle f \uparrow g, k \rangle = \sum_{h \in A^*} (-1)^{|h| + |k|} \binom{k}{h}\binom{h}{f}\binom{h}{g}. \qquad \blacksquare$$

Remark 6.3.21. If f and g are restricted to be words over a one-letter alphabet, (6.3.25) reduces to an expansion formula for the classical binomial coefficients. More precisely let p and q be positive integers with $p \geqslant q$. Then

$$a^p \uparrow a^q = \sum_{s=p}^{p+q} \binom{s}{p}\binom{p}{s-q} a^s,$$

and (6.3.25) becomes

$$\forall p, q, r \in \mathbb{N}, \qquad p \geqslant q, \qquad \binom{r}{p}\binom{r}{q} = \sum_{s=p}^{p+q} \binom{s}{p}\binom{p}{s-q}\binom{r}{s},$$

a recorded combinatorial identity (see Riordan 1968:15).

Conversely, relation (6.3.25) gives a complete set of finite identities for the Magnus expansion of an element of A^*. Once it is noted that all the coefficients of $\mu(f)$, for f in A^*, are positive, this converse may be stated as follows:

THEOREM 6.3.22. *Let s be a nonzero element of $\mathbb{N}\langle\langle A \rangle\rangle$ such that the following holds:*

$$\forall f, g \in A^*, \qquad \langle s, f \rangle\langle s, g \rangle = \sum_{w \in A^*} \langle f \uparrow g, w \rangle\langle s, w \rangle \quad (6.3.29)$$

Then there exists a unique word h in A^* such that $s = \mu(h)$; that is, for all f in A^*, $\langle s, f \rangle = \binom{h}{f}$.

Proof. Some notations prove to be useful. Let $k = \text{Card}(A)$ and let $A = \{a_1, a_2, \ldots, a_k\}$. Let α be the surjective homomorphism from A^* onto $\mathbb{N}^k$ defined by: $\alpha(a_i)$ is the element of $\mathbb{N}^k$, all the coordinates of which are 0 but the ith one, which is 1. For all f in A^* we thus have

$$\alpha(f) = \left(\binom{f}{a_1}, \binom{f}{a_2}, \ldots, \binom{f}{a_k} \right).$$

Let $\leqslant$ denote the natural (partial) ordering of $\mathbb{N}^k$. That ordering is the image under α of the division ordering on A^*; that is,

$$\forall f, g \in A^*, \qquad \binom{f}{g} \neq 0 \Rightarrow \alpha(g) \leqslant \alpha(f). \tag{6.3.30}$$

It is easy to establish by induction that the following holds:

$$\forall f, g, w \in A^*, \qquad \langle f \uparrow g, w \rangle \neq 0 \Rightarrow \alpha(f) \leqslant \alpha(w),$$
$$\alpha(g) \leqslant \alpha(w), \quad \text{and} \quad \alpha(w) \leqslant \alpha(f) + \alpha(g) \tag{6.3.31}$$

Finally let $S = (\langle s, a_1 \rangle, \langle s, a_2 \rangle, \ldots, \langle s, a_k \rangle)$.

If we take g to be a letter of A in (6.3.29), we have, using Lemma 6.3.16 and relations (6.3.30) and (6.3.31):

$$\forall f \in A^*, \qquad \forall a \in A, \qquad \langle s, f \rangle \left[\langle s, a \rangle - \binom{f}{a} \right] = \sum_{\substack{w \in A^* \\ \alpha(w) = \alpha(f) + \alpha(a)}} \binom{w}{f} \langle s, w \rangle \tag{6.3.32}$$

From now on we make constant and implicit use of the fact that all the $\langle s, f \rangle$ as well as all the binomial coefficients are positive integers. The first consequence of (6.3.32) is then:

Claim 1. Let T be an element of $\mathbb{N}^k$ that is not smaller than or equal to S; that is, $S - T$ is an element of $\mathbb{Z}^k \setminus \mathbb{N}^k$. Then for all f such that $\alpha(f) = T$, $\langle s, f \rangle = 0$.

In particular if S is 0 (of $\mathbb{N}^k$), $\langle s, f \rangle = 0$ for all f in A^+ and $\langle s, 1 \rangle = 1$ by taking $f = g = 1$ in (6.3.29) and using the assumption that s is nonzero. We are done with $h = 1$. Let suppose now that S is not the zero element of $\mathbb{N}^k$.

Claim 2. For f in A^* such that $\alpha(f) = S$, then $\langle s, f \rangle$ equals 0 or 1.

Indeed if we make $f = g$ in (6.3.29) using (6.3.21), (6.3.31), and Claim 1 we have:

$$\langle s, f \rangle^2 = \sum_{w \in A^*} \langle f \uparrow f, w \rangle \langle s, w \rangle = \langle s, f \rangle.$$

Claim 3. There exists a unique h in A^* such that $\alpha(h) = S$ and $\langle s, h \rangle = 1$.

This is because, if for all f such that $\alpha(f) = S$ we had $\langle s, f \rangle = 0$, a repetitive use of (6.3.32) would give that for all v such that $\alpha(v) < S, \langle s, v \rangle$ is also equal to 0. This is in contradiction with the assumption that s is different from 0. On the other hand if there were two distinct words, say f and g, such that $\alpha(f) = \alpha(g) = S$ and $\langle s, f \rangle = \langle s, g \rangle = 1$, (6.3.29) would read

$$\sum_{w \in A^*} \langle f \uparrow g, w \rangle \langle s, w \rangle = 1,$$

which contradicts the conjunction of Theorem 6.3.18 and of Claim 1: the first one asserts that $f \uparrow g$ is a polynomial of valuation strictly greater than $|f| = |g|$ since f and g are distinct, and the second one ensures that $\langle s, w \rangle = 0$ if w is of length strictly greater than $|f|$. Now the value of $\langle s, f \rangle$ is determined for all f such that $\alpha(f) = S$. A repetitive use of (6.3.32) uniquely determines the value of $\langle s, f \rangle$ for all f such that $\alpha(f) < S$. But the numbers $\binom{h}{f}$, for all f, also fulfill (6.3.32) and coincide with $\langle s, f \rangle$ for all f such that $\alpha(f) = S$; they must coincide for all f such that $\alpha(f) < S$. By Claim 1 and (6.3.30) they also coincide for all f such that $\alpha(f) \nleq S$. ∎

Like Theorem 6.3.12, Theorem 6.3.14 is due to Chen, Fox, and Lyndon (1958). But it was stated there in such a context—namely the number $\langle s, f \rangle$ needed not to be positive any more—that its proof was of a completely different nature. Both theorems, as well as the definition of the infiltration, are given also in Ochsenschläger 1981a independently of the first reference, and in the restricted domain of the binomial coefficients. Our presentation, and especially the proof of Theorem 6.3.14, is modeled after a preliminary version of this latter reference.

As stated several times in this section and once again just before, binomial coefficients may be considered in a more general and algebraic framework, namely the Magnus representation of the free group and the free differential calculus of Fox. This aspect will be sketched in the problems. It leads also to relations between the shuffle product and the free Lie algebra that were mentioned in the problems of Chapter 5.

Problems

Section 6.1

6.1.1. On A^* define $f \leqslant g$ if f is a *factor* of g. Show that $\leqslant$ is a partial ordering of A^* for which there exist infinite sets of pairwise noncomparable elements.

6.1.2. *Well partial ordering.* Prove the following corollaries of Proposition 6.1.3:

    a. Any subset of a well partially ordered (WPO) set is a WPO set.

    b. Let E and F be two partially ordered (PO) sets. A mapping α from E into F is a homomorphism if for any x and y in E $x \leqslant y$ implies $\alpha(x) \leqslant \alpha(y)$. Prove that any homomorphic image of a WPO set is a WPO set.

    c. Let E and F be two PO sets; the product $E \times F$ is canonically ordered by $(x, t) \leqslant (y, u)$ iff $x \leqslant y$ and $t \leqslant u$ for all x, y in E and all t, u in F. The definition extends to the product of any finite number of PO sets. Prove that the product of any finite number of WPO sets is a WPO set.

    d. The free commutative monoid $\mathbb{N}^k$ is thus canonically partially ordered by the natural ordering on each of its components. Prove, in two different ways, the theorem of Dickson (1903): $\mathbb{N}^k$ *is a WPO set.*

6.1.3. For any subset X of A^* let $(\overline{X})_c$ denote the complement of $\overline{X}$, the shuffle ideal generated by X; that is, $(\overline{X})_c = \{ f \mid \forall x \in X \quad x \nmid f \}$.

    a. Show that for f and g in A^* and for a and b in A,

$$\left(\overline{fa} \right)_c \left(\overline{bg} \right)_c = \left(\overline{fabg} \right)_c \cup \delta_{a,b} \left(\overline{fa} \right)_c a \left(\overline{bg} \right)_c.$$

    b. Deduce Theorem 6.1.2 from part a and by induction on Card(A). (See Jullien 1968.)

6.1.4. *Antichains.* Let E be a well partially ordered set. For convenience, we call *antichain* a set of pairwise incomparable elements. Certain antichains are contained in arbitrarily large antichains, we call them *narrow*. Others that do not have this property we call *wide*.

    In A^*, with the division ordering, the set $X = \{aba, bab\}$ is narrow since $X \cup A_n \cup B_n$ is an antichain for every integer n, with $A_n = \{a^i b^{n-i} \mid i < n\}$ and $B_n = \{b^i a^{n-i} \mid i < n\}$, while $Y = \{ab, ba\}$ is wide.

    a. Consider first $\mathbb{N}^k$, with the natural ordering. Show that any antichain of $\mathbb{N}^2$ is wide; moreover, given an antichain $X = \{(x_1, y_1), (x_2, y_2), \ldots, (x_p, y_p)\}$ of $\mathbb{N}^2$, compute the maximum of the cardinal of the antichains which contain X. For $k \geqslant 3$, give a necessary and sufficient condition for an antichain of $\mathbb{N}^k$ to be wide.

    b. Let $k = \text{Card}(A)$ and let α be a surjective homomorphism from A^* onto $\mathbb{N}^k$. Give a necessary and sufficient condition for an antichain X of $\mathbb{N}^k$ to be such that $\alpha^{-1}(X)$ is wide. See Problem 6.2.9 for a characterization of wide antichains of A^*.

6.1.5. *Maximal chains.* A chain is a subset of A^* totally ordered by division. A chain $\{f_0, f_1, \ldots, f_n\}$ is said to be maximal if it is maximal among the chains the last element of which is f_n (we make the implicit assumption that $f_i \mid f_j$ iff $i \leqslant j$). A maximal chain always begins with the empty word, and its length (that is, the number of elements) is $|f_n| + 1$. We denote by $\mathcal{C}_n$ the set of maximal chains of length n. We suppose that $A = \{a, b\}$ is a two-letter alphabet.

a. Let $f = x_1 x_2 \cdots x_p$ be a word in A^* of length p. Define the words $g_0, g_1, \ldots, g_{p+1}$ by the following: $g_0 = bf$, $g_{p+1} = fa$, and $g_i = x_1 \cdots x_{i-1} ab x_{i+1} \cdots x_p$ for $1 \leqslant i \leqslant p$. Show that these words are distinct and are exactly the $(p+2)$ words of length $p+1$, having f as subword. Deduce that $\text{Card}(\mathcal{C}_{n-1}) = n!$

b. Let σ be a permutation of the set $[n] = \{1, 2, \ldots, n\}$ and let σ be represented as a word over $[n]$; that is, $\sigma = \sigma(1)\sigma(2) \cdots \sigma(n)$. Recall then that the *up–down sequence* of σ is the word f of A^* of length $n-1$, $f = x_1 x_2 \cdots x_n$ defined by $x_i = a$ if $\sigma(i) < \sigma(i+1)$, and $x_i = b$ in the opposite case. Let $\alpha = \{f_0, f_1, \ldots, f_{n+1}\}$ be in $\mathcal{C}_{n-1}$ and define the sequence of permutations σ_p, $1 \leqslant p \leqslant n$, by the following induction: $\sigma_1 = 1$, and, for $1 \leqslant p \leqslant n, \sigma_{p+1} = z_1 z_2 \cdots z_i(p+1)z_{i+1} \cdots z_p$ if $\sigma_p = z_1 z_2 \cdots z_p$ and if f_p is numbered by i in the process described in problem part a giving the $(p+1)$ words of length p having f_{p-1} as subwords; define $\Phi(\alpha) = \sigma_n$. Show that σ_{p+1} is the unique permutation obtained by inserting the letter $(p+1)$ in σ_p and having f_p as up–down sequence. Deduce that Φ is a bijection between $\mathcal{C}_{n-1}$ and $\mathcal{G}_n$, the set of permutations on n elements, and derive an algorithm to compute the number $p(f)$ of permutations having f as up–down sequence.

6.1.6. *Maximal chains* (continued). The cardinality of A is now any positive integer k.

a. Let $f = x_1 x_2 \cdots x_p$ be a word of A^* of length p. Define the set of words $G_0, G_1, \ldots, G_p$ by the following: $G_0 = Af$, and $G_i = x_1 x_2 \cdots x_i (A \setminus x_i) x_{i+1} \cdots x_p$ for $1 \leqslant i \leqslant p$. Show that the sets G_i are pairwise disjoint and that $G = \cup_0^p G_i$ is the set of the $((p+1)(k-1)+1)$ words of length $p+1$ that have f as subword. Deduce from this an expression for $\text{Card}(\mathcal{C}_{n-1})$.

b. Let $\gamma = \{f_0, f_1, \ldots, f_n\}$ be a maximal chain and define *the trace of* γ to be the word $\text{Tr}(\gamma) = y_1 y_2 \cdots y_n$ where y_i is the letter that must be cancelled from f_i in order to get f_{i-1}. Let $w = y_1 y_2 \cdots y_n$ be in A^*; for each $j, 1 \leqslant j \leqslant n$ define $l(w, j)$ to be the number of occurrences of letters, different from y_j, occurring in $y_1 y_2 \ldots y_j$, increased by 1. Let

$$Fo(w) = \prod_{j=1}^{n} l(w, j).$$

Deduce from part a that the number of maximal chains γ such that $\text{Tr}(l) = w$ is equal to $Fo(w)$.

The reference for both Problems 6.1.5 and 6.1.6 is Viennot 1978. As in Problem 6.1.5 the process described in Problem 6.1.6a gives rise, for $k = 2$, to a bijection between $\mathcal{C}_{n-1}$ and $\mathcal{G}_n$, but its description becomes more involved.

6.1.7. For every f in A^* define $\beta(f)$ to be the length of the longest subword of f in which any two consecutive letters are distinct; for instance $\beta(abbaaabaabb) = 6$. Show that there are exactly $\beta(f)$ distinct subwords of f, of length $|f| - 1$.

6.1.8. *Möbius function.* For all f and g in A^* define $\mu(f, g)$ by the following: $\mu(f, g) = 0$ if $f \nmid g$, $\mu(f, f) = 1$ for all f, and $\mu(f, g) = -\Sigma\mu(f, h)$ where the sum ranges over the words h such that $f | h, h | g$, and $h \neq g$ (thus the sum is finite).

a. Let s and t be two real-valued functions on A^*. Show that $s(f) = \Sigma_{g|f} t(g)$ if and only if $t(g) = \Sigma_{f|g} s(f)\mu(f, g)$. *Hint:* Define $\zeta(f, g)$ by $\zeta(f, g) = 1$ if $f | g$, and $\zeta(f, g) = 0$ otherwise and show that

$$\delta_{f,g} = \sum_{h \in A^*} \zeta(f, h)\mu(h, g).$$

b. For f in A^*, let $\beta(f)$ be as in exercise 1.10: Show that $\mu(1_{A^*}, f) = (-1)^{|f|}$ if $\beta(f) = f$ and $\mu(1_{A^*}, f) = 0$ otherwise.

Möbius functions on arbitrary partially ordered sets are defined and studied in Rota 1964. Part b is a remark in Viennot 1978.

6.1.9. *Shuffle and infiltration product.* In order to distinguish between the sub-sequences of a word and the subwords they define, let us take the following notation: For every integer n, $[n] = \{1, 2, \ldots, n\}$; if h is a word and I a subset of $[|h|]$ then $h_I = h_{i_1} h_{i_2} \cdots h_{i_k}$ where $i_1 < i_2 < \cdots < i_k$ and $I = \{i_1, i_2, \ldots, i_k\}$.

a. Show that a word h belongs to the shuffle of two words f and g if and only if there exists a pair (I, J) of subsets of $[|h|]$ such that
  (i) $h_I = f$, $\quad h_J = g$.
  (ii) $I \cup J = [|h|]$.
  (iii) $I \cap J = \varnothing$.

b. Define the infiltration product of two words f and g to be the set of words h such that there exists a pair (I, J) of subsets of $[|h|]$ that satisfies conditions (i) and (ii) in part a. Denote the infiltration of f and g by $f \uparrow g$. Show that for all f and g in $A^* \bar{f} \cap \bar{g} = \overline{f \uparrow g}$. If val$(f \uparrow g)$ denote the set of words of minimal length of $f \uparrow g$, verify that $\overline{f \uparrow g}$ is not always included in val$(f \uparrow g)$.

Section 6.2

6.2.1. Let f be a word in A^* and $m = |f|$. Of course $[f]_{m+1}$ always consists of f alone. The class $[f]_m$ needs not to be a singleton since $[a^m]_m = a^m a^*$ with a in A; show that this is the only possibility for $[f]_m$ not to be a singleton. On the other one hand, there exist words f such that $[f]_n$ is a singleton with $n < m$; e.g., $|[abab]_3| = 1$. Give an algorithm that, given any word f, computes efficiently the smallest integer n such that $[f]_n$ is a singleton.

6.2.2. Let A be the two-letter alphabet $\{a, b\}$ and let m be a positive integer. Let f and g be in A^*. Show that $af \equiv bg[J_m]$ if, and only if $S(m, af) = S(m, bg) = A^{\leqslant m}$.

6.2.3. Let A be the two-letter alphabet $\{a, b\}$ and let m be a positive integer. Let f and g be in A^*. Show that $abf \equiv abg[J_m]$ if, and only if, $f \equiv g[J_{m-1}]$.

6.2.4. Based on the ideas in the proof of Theorem 6.2.11 and of Proposition 6.2.15, write an efficient algorithm to find a shortest word that distinguishes two given distinct words (in $O(|fg|^3)$ steps, where f and g are the given words).

6.2.5. Let R_m be the relation on A^* defined by $fagR_m fg$ iff $f, g \in A^*$, $a \in A$ and $fg \equiv fag[J_m]$. Show that R_m^* is the intersection of the relation of division with the equivalence relation J_m. Show that $J_m = R_m^{-1*} \circ R_m^*$ $= (R_m \cup R_m^{-1})^*$.

6.2.6. Given a word f in A^* and an integer $m > 0$, prove that $[f]_m$ is a singleton if and only if for every factorization $f = gh$ of f and for every letter a in A, $\delta(gah, gh) < m$.

6.2.7. Given a word f in A^* and an integer $m \geqslant 0$, prove that f is a minimal element of $[f]_m$ if and only if for every g and h in A^* and for every a in A, if $f = gah$ then $\delta(gah, gh) < m$.

6.2.8. Given a word f in A^* and an integer $m \geqslant 0$, show that if $[f]_m$ contains an antichain of cardinality two then it also contains antichains of arbitrarily large cardinality. As a consequence, each class modulo J_m is either a chain or contains arbitrarily large antichains.

6.2.9. Show that an antichain X of A^* is wide (see Problem 6.1.4) if and only if $(\bar{X})_c$ (see Problem 6.1.3) is the union of a finite family of chains.

6.2.10. Two elements m and m' of a monoid M are said to be $\mathcal{J}$-equivalent if they generate the same two-sided ideal (that is, $m \mathcal{J} m' \Leftrightarrow MmM = Mm'M$). A monoid M is said to be $\mathcal{J}$-trivial if the equivalence $\mathcal{J}$ is the identity.

   a. Show that the quotient of A^* by any congruence J_m is $\mathcal{J}$-trivial. (*Hint:* Use Corollary 6.2.16 and its dual.)

   b. Let M be a finite $\mathcal{J}$-trivial monoid. Show that there exist a free monoid A^* and an integer m such that M is a homomorphic image of the quotient of A^* by J_m. (*Hint:* (i) Let $k = \text{Card}(M)$ and γ a surjective homomorphism from a suitable free monoid A^* onto M. Using Proposition 6.2.15, show that $g \equiv ag[J_k]$ implies that $\gamma(g) = \gamma(ag)$. (ii) Let $m = 2k$. Using Lemma 6.2.10 show that $fag \equiv fg[J_m]$ implies $\gamma(fag) = \gamma(fg)$; complete the proof with Problem 6.2.5.) (See Simon 1975; for an alternative proof see Eilenberg 1976.)

6.2.11. Let f be a word of A^* and let $V(f)$ be the set of the *factors* of f of length $[(f/2) + 1]$. Show that if f is not of the form $f = (uv)^k u$ with

$k \geqslant 2$ then f is uniquely determined by $V(f)$. Verify that f would not be determined if the length of the factors in $V(f)$ were $[f/2]$.

6.2.12. Let f be a word in A^* and let m be an integer, such that $m \geqslant (|f| + 1)/2$. Show that the class $[f]_m$ is a chain, relative to the division ordering. Based on Proposition 6.2.20 find a better bound for m.

6.2.13. Characterize the pairs (f, m), f in A^* and m in $\mathbb{N}$ for which the class $[f]_m$ is a chain, relative to the division ordering.

Section 6.3

6.3.1. a. Show that for all f in A^* $\sum_{g \in A^*} \binom{f}{g} = 2^{|f|}$

   b. Show that for all f and g in A^* $\sum_{h \in A^*} \langle f \circ g, h \rangle = \binom{|f| + |g|}{|g|}$

6.3.2. a. Prove that for all f and g in A^*, and all a in A one has

$$\binom{f}{ga} = \sum_{\substack{u \in A^* \\ uav = f}} \binom{u}{g}$$

   (*Hint*: Use an argument similar to the one in the proof of Proposition 6.3.2.)

   b. Show, without any induction argument, that for all g in A^* and all a in A the following holds:

$$ga \circ A^* = (g \circ A^*)aA^*$$

   c. Deduce Proposition 6.3.13 from parts a and b.

6.3.3. Let M and P be two matrices over the same semi-ring of coefficients, of dimension m, n, and p, q, respectively. As usual the tensor product of M by P, denoted by $M \otimes P$ is the matrix of dimension mp, nq, the entries of which are given by

$$M \otimes P_{ip+k, jq+l} = M_{i,j} P_{k,l}.$$

Tensor product is associative.

   Let A be an alphabet of cardinality k, I the identity matrix of dimension k, and U the column vector of dimension k, all entries of which are 1.

   a. Let V be the column vector (resp. the row vector) of dimension k, the entries of which are the letters of A, in a given total order $\leqslant$. Show that $V \otimes V \otimes \cdots \otimes V$, where the product has n factors, is the column vector (resp. the row vector) of dimension k^n, the entries of which are the words of length n, in the lexicographic ordering induced by $\leqslant$.

   b. Let C_q^p be the (k^p, k^q)-matrix over the integers, the rows of which are indexed by A^p, in the lexicographic ordering, the columns of which are indexed by A^q, in the lexicographic ordering, and such that the entry (f, g), with f in A^p and g in A^q, is equal to $\begin{pmatrix} f \\ g \end{pmatrix}$. Show that

$$C_q^p = C_q^{p-1} \otimes U + C_{q-1}^{p-1} \otimes I.$$

6.3.4. *Magnus representation of free groups.* Let A be an alphabet and F the free group generated by A.

a. Show that the homomorphism $\mu: A^* \to \mathbb{Z}\langle A^* \rangle$, defined by $\mu(a) = 1 + a$ for a in A, extends to an injective homomorphism of F into $\mathbb{Z}\langle\langle A^* \rangle\rangle$. (*Hint:* $\mu(a^{-1}) = 1 - a + a^2 - \cdots + (-1)^n a^n + \cdots$)

b. For u, v in F let us denote by $[u, v] = u^{-1}v^{-1}uv$ the commutator of u and v. Let F_n be the nth term of the lower central series of F; that is, $F_0 = F$, $F_{n+1} = [F_n, F]$ where $[U, V]$ is the subgroup generated by the elements $[u, v]$, $u \in U$, $v \in V$. Let D_n be the subgroup formed by the elements w of F such that $\mu(w) - 1$ is a series of valuation $\geq n$ (D_n is called the nth dimension subgroup). Show that $F_n \subseteq D_n$. Indeed we have $F_n = D_n$ (See Problem 5.3.9). Show that two elements u and v of F (and thus of A^*) are such that $\mu(u) - \mu(v)$ is a series of valuation $\geq n$ if and only if uv^{-1} belongs to F_n. (See Magnus, Karass, and Solitar 1976.)

c. Let $(u_n)_{n \in \mathbb{N}}$ and $(v_n)_{n \in \mathbb{N}}$ be two sequences of words in $\{a, b\}^*$ defined by

$$u_0 = a, \qquad v_0 = b,$$
$$u_{n+1} = u_n v_n, \qquad v_{n+1} = v_n u_n.$$

These are the Morse–Thue sequences of Chapter 2. Show that for any f of length less that n, $\begin{pmatrix} u_n \\ f \end{pmatrix} = \begin{pmatrix} v_n \\ f \end{pmatrix}$. (*Hint:* Use part b and Proposition 6.3.6.)

d. Show that there exists an f of length n such that $\begin{pmatrix} u_n \\ f \end{pmatrix} \neq \begin{pmatrix} v_n \\ f \end{pmatrix}$. (See Ochsenschläger 1981b.)

6.3.5. *Free differential calculus.* Let F be the free group generated by A and let ε be the homomorphism (of $\mathbb{Z}$-algebra) from $\mathbb{Z}\langle F \rangle$ into $\mathbb{Z}$ induced by the trivial map $\varepsilon: F \to 1$; then for each s in $\mathbb{Z}\langle F \rangle$, $s = \sum_{w \in F} \langle s, w \rangle w$ where the integer $\langle s, w \rangle$ is zero for all but a finite number of w, $\varepsilon(s) = \sum_{w \in F} \langle s, w \rangle$. For each a in A let $\partial/\partial a$, the *derivation with respect to a*, be the mapping of $\mathbb{Z}\langle F \rangle$ into itself defined by

$$\forall a, b \in A, \qquad \frac{\partial}{\partial a} b = \delta_{a, b} \tag{1}$$

$$\forall u, v \in F, \qquad \frac{\partial}{\partial a}(uv) = \frac{\partial}{\partial a}u + u\frac{\partial}{\partial a}v \tag{2}$$

$$\forall s, t \in \mathbb{Z}\langle F \rangle, \qquad \frac{\partial}{\partial a}(s + t) = \frac{\partial}{\partial a}s + \frac{\partial}{\partial a}t \tag{3}$$

that is, (3) expresses that $\partial/\partial a$ is an homomorphism of $\mathbb{Z}$-module.

a. Prove that for all s and t in $\mathbb{Z}\langle F\rangle$

$$\frac{\partial}{\partial a}(st) = \left(\frac{\partial}{\partial a}s\right)\varepsilon(t) + s\frac{\partial}{\partial a}t. \tag{4}$$

b. Prove that for all s in $\mathbb{Z}\langle F\rangle$ the following fundamental formula holds:

$$s = \varepsilon(s) + \sum_{a\in A}\left(\frac{\partial}{\partial a}s\right)(a-1).$$

c. For every f in A^* define the derivation with respect to f by the following induction:

$$\frac{\partial}{\partial af} = \frac{\partial}{\partial a}\circ\frac{\partial}{\partial f}.$$

Prove that for all w in F and all f in A^* one has

$$\varepsilon\left(\frac{\partial}{\partial f}w\right) = \langle\mu(w), f\rangle.$$

(See Fox, 1953.)

**d. Prove that (6.3.25) holds again when h is taken to be any element of F (and if $\binom{h}{f}$ is understood to be defined as $\langle\mu(h), f\rangle$). (See Chen, Fox, and Lyndon 1958.)

6.3.6. For any integer q, let $\mathbb{K}_q$ be the quotient of $\mathbb{N}$ by the congruence generated by $q = q+1$ and let ϕ_q be the canonical homomorphism from $\mathbb{N}$ onto $\mathbb{K}_q$; ϕ_q extends to a homomorphism from $\mathbb{N}\langle A\rangle$ onto $\mathbb{K}_q\langle A^*\rangle$. For every integer n let $I_{q,n}$ be the congruence of A^* defined by

$$f \equiv g\lfloor I_{q,n}\rfloor \;\Leftrightarrow\; \mathrm{val}\big(\phi_q(\mu(f-g))\big) > n.$$

Show that the classes of $I_{q,n}$ belong to the boolean algebra of shuffle ideals. (See Eilenberg 1976.)

6.3.7. For any prime integer p let ψ_p be the canonical homomorphism from $\mathbb{N}$ onto $\mathbb{Z}_p = \mathbb{Z}/p\mathbb{Z}$; ψ_p extends to a homomorphism from $\mathbb{N}\langle A\rangle$ onto $\mathbb{Z}_p\langle A\rangle$. For every integer n let $H_{p,n}$ be the congruence on A^* defined by

$$f \equiv g[H_{p,n}] \;\Leftrightarrow\; \mathrm{val}\big(\psi_p(\mu(f-g))\big) > n$$

Show that the quotient of A^* by $H_{p,n}$ is a finite p-group $G_{p,n}$ and that, conversely, for any finite p group G there exists an n such that G is a quotient of $G_{p,n}$. (See Eilenberg 1976.)

6.3.8. *Iterated integrals of Chen.* Let $\mathbb{R}^m$ be an affine m-space with the coordinate $(x_1, x_2, \ldots, x_m)$. Let a *path* α be a function, $[x, y] \to \mathbb{R}^m$, $\alpha(t) = (\alpha_1(t), \alpha_2(t), \ldots, \alpha_m(t))$, such that each $\alpha_i(t)$ is continuous and of bounded variations for $x \leqslant t \leqslant y$. Let $A = \{a_1, a_2, \ldots, a_m\}$ be an alphabet of cardinality m; to each word f of A^*, and to each path α we associate an iterated integral symbolically denoted by $\int_\alpha df$, and inductively defined by

$$\forall a_i \in A, \qquad \int_\alpha da_i = \alpha_i(y) - \alpha_i(x) = \int_x^y d\alpha_i(t),$$

$$\forall f \in A^*, \qquad \forall a_i \in A, \qquad \int_\alpha df\, a_i = \int_x^y \left(\int_x^t df \right) d\alpha_i(t).$$

If P is a polynomial $\int_\alpha dP$ is defined by linearity. Show that $(\int_\alpha df)(\int_\alpha dg) = \int_\alpha d(f \circ g)$. (See Chen 1957; Ree 1958.) This correspondence between the product of iterated integrals and the shuffle of words has been also observed by Fliess (1981). This last author interprets a formal power series $s = \sum_{f \in A} \langle s, f \rangle f$ of $\mathbb{R}\langle\langle A \rangle\rangle$ as a functional S over the path α of $\mathbb{R}^m$ by means of the formula

$$S(\alpha) = \sum_{f \in A^*} \langle s, f \rangle \int_\alpha df.$$

Within this framework the addition of functionals corresponds to the addition of series; show that the product of functionals corresponds to the shuffle product of series.

6.3.9. *Shuffle and infiltration product.* Take the same notations as in Problem 6.1.12.

a. Show that for any f, g, and h in A^*, $\langle f \circ g, h \rangle$ is equal to the number of distinct pairs (I, J) such that
   (i) $h_I = f$ and $h_J = g$.
   (ii) $I \cup J = [|h|]$.
   (iii) $I \cap J = \varnothing$.

b. Show that $\langle f \uparrow g, h \rangle$ is equal to the number of distinct pairs (I, J) that satisfy conditions (i) and (ii).

CHAPTER 7

Unavoidable Regularities in Words and Algebras with Polynomial Identities

7.0. Introduction

This chapter presents combinatorial results, regarding unavoidable regularities in words, close to Ramsey's and van der Waerden's theorems (see Chapters 3 and 4) and applications to algebras with polynomial identities, concerning the Kurosch, Burnside, and Levitzki problem (the Kurosch problem is to know whether a finitely generated algebraic algebra is finite dimensional). Of course, the point of view here is combinatorial and not ring-theoretic; that is why we give all the definitions of pi-algebras and elementary properties that we need. What is interesting in this combinatorial approach is that it enables us to obtain quickly, without any deep knowledge of ring theory, the beautiful theorem of Shirshov on pi-algebras.

In the first section some combinatorial results are proved (the principal one of which is Theorem 7.1.5), which is used in Section 7.2 to prove Shirshov's theorem on pi-algebras (Theorem 7.2.2). The third section gives some corollaries to that theorem that are of independent interest.

7.1. Some Combinatorial Results

Consider a totally ordered alphabet A with a smallest element denoted by a: $a = \min(A)$. The notation $\underset{A}{\leqslant}$ is used here to denote the lexicographic order on A^* (see Chapter 5, Section 5.1). This notation will allow us to consider several alphabets.

The subset X of A^* defined by

$$X = a^+ (A - a)^+$$

is a code (that is, the basis of a free submonoid, see Chapter 1), as is easily verified. The submonoid X^* is the set of words beginning by a and ending with another letter than a, together with the empty word:

$$X^* = 1 \cup aA^*(A - a)$$

X can be considered as an alphabet, totally ordered by $\underset{A}{\leqslant}$; one defines then the lexicographic order $\underset{X}{\leqslant}$ on X^* as just shown.

The following lemma shows that in X^*, the orders $\underset{A}{\leqslant}$ and $\underset{X}{\leqslant}$ are identical.

LEMMA 7.1.1. *Let $f, g \in X^*$ such that $f \underset{X}{\leqslant} g$. Then $f \underset{A}{<} g$*

Proof. (We must point out that it is not superfluous.) If f is a left factor of g, it is clear. Otherwise $f = uxv$, $g = uyw$ for some words $u, v, w \in X^*$ and $x, y \in X$ such that $x < y$. The following explains that last inequality:

Either $x = rbs$, $y = rct$ with $r, s, t \in A^*$ and $b, c \in A$ such that $b < c$. Then $f = (ur)b(sv)$ and $g = (ur)c(tw)$; hence $f \underset{A}{<} g$. Or x is a proper left factor of y: $y = xbt$ with $b \in A$ and $t \in A^*$. From the definition of X it follows that $b \neq a$, hence $a < b$. Then $f = uxv$, $g = uxbtw$. If $v = 1$, f is a proper left factor of g, hence $f \underset{A}{<} g$. If $v \neq 1$, because $v \in X^*$ one has $v = av'$, hence $f = (ux)av'$, $g = (ux)b(tw)$, and $f \underset{A}{<} g$. ∎

A word $w \in A^*$ is *n-A-divided* if there exist words $w_1, w_2, \ldots, w_n$ in A^+ such that $w = w_1 w_2 \cdots w_n$ and that for any permutation $\sigma \in \mathfrak{S}_n - id$

$$w \underset{A}{<} w_{\sigma(1)} w_{\sigma(2)} \cdots w_{\sigma(n)}.$$

There is a formal analogy between this definition and the one of the Lyndon words (see Chapter 5). However, we must beware of misleading confusions. We shall speak, in the same way, of words in X^* that are *n-X-divided*. The following lemma allows us mechanically to associate to any word in X^* that is $(n-1)$-X-divided a word in A^* that is n-A-divided. In the sequel it will allow recursion arguments.

LEMMA 7.1.2. *If $w \in X^*$ is $(n-1)$-X-divided then wa is n-A-divided.*

Example 7.1.3. Let $x, y \in X^*$ such that $xy \underset{X}{<} yx$. By definition xy is 2-X-divided. Then the word xya in A admits the 3-A-division $xya = a(x'a)(y'a)$ where $x = ax'$ and $y = ay'$. We verify it partially for the two permutations (23) and (123). The first one gives the word $ay'ax'a = yxa \underset{A}{<} xya$. The second one gives the word $x'ay'aa = x'yaa > ax'ya$ because

$x' = a^i bt$ with $i \geq 0$, $b > a$; hence x' is strictly greater than $ax' = a^i abt$ in the order $\underset{A}{\leq}$.

Proof. w can be written as $w = w_1 w_2 \cdots w_{n-1}$ where the w_is in X^+ are such that for any permutation $\sigma \in \mathfrak{S}_{n-1} - id$, $w \underset{X}{<} w_{\sigma(1)} \cdots w_{\sigma(n-1)}$. By definition of X, we have $w_i = aw_i'$, $i = 1, \ldots, n-1$. Let $u_1 = a$, $u_2 = w_1' a, \ldots, u_n = w_{n-1}' a$. Then $wa = u_1 u_2 \cdots u_n$.

We verify that this factorization of wa is an n-A-division.

For each $\sigma \in \mathfrak{S}_n$, let $\bar{\sigma} = u_{\sigma(1)} \cdots u_{\sigma(n)}$. There exist α, τ in $\mathfrak{S}_n$ such that $\sigma = \alpha \circ \tau$, $\alpha(1) = 1$ and that τ is a cycle of the form $(1, \ldots, r)$. Indeed let $r = \sigma^{-1}(1)$, $\tau = (1, \ldots, r)$ and $\alpha = \sigma \circ \tau^{-1}$; then $\sigma = \alpha \circ \tau$ and $\alpha(1) = \sigma \circ \tau^{-1}(1) = \sigma(r) = 1$. Note that $\sigma \neq id$ implies that α or $\tau \neq id$.

(i) We have

$$\bar{\alpha} = au_{\alpha(2)} \cdots u_{\alpha(n)}$$
$$= aw_{\alpha(2)-1}' a \cdots w_{\alpha(n)-1}' a$$
$$= w_{\beta(1)} \cdots w_{\beta(n-1)} a$$

where $\beta \in \mathfrak{S}_{n-1}$ is defined by $\beta(i) = \alpha(i+1) - 1$. Then $\alpha \neq id$ implies $\beta \neq id$, hence by hypothesis $w \underset{X}{<} w_{\beta(1)} \cdots w_{\beta(n-1)}$. It follows by Lemma 7.1.1 that $wa \underset{A}{<} \bar{\alpha}$.

(ii) The foregoing above allows us to conclude when $\tau = id$ (because $\sigma = \alpha$ implies $\bar{\sigma} = \bar{\alpha}$). Suppose now $\tau \neq id$, that is, $r \geq 2$. We have

$$\bar{\sigma} = u_{\sigma(1)} u_{\sigma(2)} \cdots u_{\sigma(n)}$$
$$= u_{\alpha(2)} \cdots u_{\alpha(r)} u_{\alpha(1)} u_{\alpha(r+1)} \cdots u_{\alpha(n)}$$
$$= u_{\alpha(2)} \cdots u_{\alpha(r)} au_{\alpha(r+1)} \cdots u_{\alpha(n)}$$

From $\alpha(2) \neq 1$ it follows that $u_{\alpha(2)} = a^k bv$ with $k \geq 0$, $b \neq a$; hence $b > a$ and $v \in A^*$, by definition of X. Hence $\bar{\sigma} = a^k bv'$ and $\bar{\alpha} = a^{k+1} bv''$ with $v', v'' \in A^*$. Hence $\bar{\alpha} < \bar{\sigma}$. If $\alpha = id$ then $wa = \bar{\alpha}$; hence $wa < \bar{\sigma}$. If $\alpha \neq id$, from (i) $wa \underset{A}{<} \bar{\alpha}$; hence $wa \underset{A}{<} \bar{\sigma}$. ∎

The following theorem regards unavoidable regularities in words. It says that given integers n and p, every long word contains as a factor either an n-divided word or a pth power. Let us recall that a factor of a word w is a word u such that $w = xuy$ for some words x, y.

THEOREM 7.1.4. *For all integers $k, p, n \geq 1$ there exists an integer $N(k, p, n)$ such that for any totally ordered alphabet A with k elements any word w in A^**

of length at least $N(k, p, n)$ *contains as a factor either a pth power of a nonempty word or an n-A-divided word.*

Proof (by double induction on n and k). Fix $p \geqslant 1$. It is clear that $N(k, p, 1)$ exists for any k. Suppose that $N(k, p, n-1)$ exists for any k ($n \geqslant 2$). The existence of $N(1, p, n)$ is clear; suppose now that $N(k-1, p, n)$ exists ($k \geqslant 2$) and let us show the existence of $N(k, p, n)$.

Let

$$N = (p + N(k-1, p, n))(N(k^{N(k-1, p, n)+p}, p, n-1) + 1)$$

Let A be a totally ordered alphabet of cardinality k and $w \in A^*$ a word of length at least N.

(i)   If w contains as a left factor a word $u \in (A \setminus a)^{N(k-1, p, n)}$, then w contains by induction a pth power or an n-divided word.

(ii) If w contains as a right factor a word $u \in a^p a^*$, then w contains a pth power.

(iii) We can therefore suppose that w contains a word w_1 with

$$|w_1| > (p + N(k-1, p, n))N(k^{N(k-1, p, n)+p}, p, n-1)$$

$$\text{and} \quad w_1 \in aA^*(A \setminus a).$$

Then

$$w_1 \in X^* \quad \text{and} \quad w_1 = x_1 x_2 \cdots x_r, x_i \in X.$$

(iv) Each x_i can be written $x_i = a^q s$ with $s \in (A \setminus a)^*$. By the same argument as in (i) and (ii), we may suppose $q < p$ and $|s| < N(k-1, p, n)$, hence $|x_i| < p + N(k-1, p, n)$. This implies

$$r > N(k^{N(k-1, p, n)+p}, p, n-1)$$

(v)   We apply the induction hypothesis to the alphabet $\{x \in X \,|\, |x| < p + N(k-1, p, n)\}$ and may conclude that $x_1 \cdots x_{r-1}$ contains a pth power or a word u that is $(n-1)$-X-divided; in this last case w_1 (hence w) contains the word ua, which is n-A-divided (Lemma 7.1.2).

We may therefore define $N(k, p, n) = N$ and the theorem is proved. ∎

It goes without saying that the numbers $N(k, p, n)$ as constructed in the last proof grow amazingly with k and n. The following result, whose proof uses a pretty combinatorial argument moderates a little bit this growth: the length of the pth power may be bounded.

THEOREM 7.1.5. *Let k, p, $n \geqslant 1$ be integers such that $p \geqslant 2n$. There exists an integer $N(k, p, n)$ such that for any totally ordered alphabet A of cardinality k, any word w in A of length at least $N(k, p, n)$ contains as a factor either an n-A-divided word or a word of the form u^p with $0 < |u| < n$.*

Proof. Let $N = N(k, p, n)$ be the integer of Theorem 7.1.4. Let A and w be as before. If w admits no n-divided factor, w contains a factor of the form v^p, $v \neq 1$. We may suppose that v is primitive (that is, v is not a power of another word, see Chapter 1, Section 1.3). If $|v| \geqslant n$, the number of conjugates of v (a conjugate of v is a word of the form fg, with $v = gf$, ibid.) is $\geqslant n$, since v is primitive. Let $v_1 < v_2 < \cdots < v_n$ be n of them. Since $p \geqslant 2n$, v^p contains as a factor the word $v^{2n} = (v^2)^n$; now v^2 contains each v_i as a factor. From that we deduce that v^p contains a word of the form

$$f = (v_1 v_1')(v_2 v_2') \cdots (v_n v_n').$$

This factorization of f is an n-division. Indeed, for each $\sigma \in \mathfrak{S}_n - id$ let i be the smallest integer such that $\sigma(i) \neq i$. Then $\sigma(1) = 1, \ldots, \sigma(i-1) = i-1, \sigma(i) = j > i$ hence σ applied to f gives the word

$$(v_1 v_1') \cdots (v_{i-1} v_{i-1}')(v_j v_j')g,$$

which is strictly greater than

$$f = (v_1 v_1') \cdots (v_{i-1} v_{i-1}')(v_i v_{i-1}')h$$

because $j > i$ implies $v_j > v_i$.   ∎

7.2. Algebras with Polynomial Identities

The results of the previous section are now used to prove a theorem regarding pi-algebras.

K denotes a commutative ring with neutral element denoted by 1. If $\mathfrak{M}$ is an (associative) K-algebra we shall always make the hypothesis that for any m in $\mathfrak{M}$ one has $1m = m$.

We denote by $K_+\langle X \rangle$ the free K-algebra without neutral element generated by X (that is, the K-algebra of noncommutative polynomials without constant term over X).

Let $\mathfrak{M}$ be a K-algebra. $\mathfrak{M}$ *satisfies the polynomial identity* $P = 0$ if there exists an alphabet $X = \{x_1, \ldots, x_q\}$, a nonzero polynomial $P(x_1, \ldots, x_q) \in K_+ \langle X \rangle$ such that for any $m_1, \ldots, m_q$ in $\mathfrak{M}$ one has $P(m_1, \ldots, m_q) = 0$. The *degree* of the identity is the degree of P. The identity is called *admissible* (according to Shirshov) if there is a word w such that $w = \deg(P)$ and that the coefficient of w in P is invertible in K (it is always true if K is a field).

The following lemma shows that a *pi-algebra* (that is, an algebra satisfying an admissible polynomial identity) always satisfies a multilinear identity.

LEMMA 7.2.1. *If $\mathfrak{M}$ satisfies an admissible polynomial identity of degree n then $\mathfrak{M}$ satisfies an identity of the form*

$$x_1 \cdots x_n = \sum_{\sigma \in \mathfrak{S}_n - id} k_\sigma x_{\sigma(1)} \cdots x_{\sigma(n)}$$

for some k_σ in K.

Proof. $\mathfrak{M}$ verifies an admissible identity of the form $P(x_1, \ldots, x_q) = 0$; we may suppose that there is a word w_0 of length $n = \deg(P)$ and of coefficient 1 in P.
(i) We may suppose that for each $x \in X$, $|w_0|_x \leq 1$. Indeed, if $|w_0|_{x_1} \geq 2$, let

$$Q = P(x_1 + x_{q+1}, x_2, \ldots, x_n) - P(x_1, \ldots, x_q) - P(x_{q+1}, x_2, \ldots, x_q)$$

where x_{q+1} is a new letter. Then $Q \neq 0$ and $\mathfrak{M}$ verifies the identity $Q = 0$. Furthermore $\deg(Q) = \deg(P)$ and, letting $w_0 = ux_1vx_1w$, the coefficient of the word $w_0' = ux_1vx_{q+1}w$ in Q is 1 and $|w_0'|_{x_1} < |w_0|_{x_1}$, $|w_0'|_{x_{q+1}} = 1$ and for each i in $\{2, \ldots, q\}$ $|w_0'|_{x_i} = |w_0|_{x_i}$. We then conclude by induction.
(ii) We may suppose $w = x_1 \cdots x_n$ and, replacing if necessary P by $P(x_1, \ldots, x_n, 0, \ldots, 0)$, that

$$X = \{x_1, \ldots, x_n\}.$$

(iii) We may suppose that $\forall w \in \text{supp}(P)$ (that is, the support of P), $\forall x \in X$, $|w|_x \geq 1$. Indeed let

$$Q = P(x_1, \ldots, x_n) - P(0, x_2, \ldots, x_n)$$

Then $\deg(P) = \deg(Q)$, $\text{supp}(Q) \subset \text{supp}(P)$, the coefficient of w_0 in Q is 1 and $Q(0, x_2, \ldots, x_n) = 0$, showing that for any w in $\text{supp}(Q)$, $|w|_{x_1} \geq 1$. We conclude by induction on the x_i's.
(iv) For each $w \in \text{supp}(P)$, we have $|w|_x \geq 1$ for each $x \in X$ by (iii).
Since $\deg(P) = n = \text{Card}(X)$, this implies $|w|_x = 1$. Finally the coefficient of $x_1 x_2 \cdots x_n$ in P is 1 and the lemma is proved. ∎

We come now to the Kurosch problem: Is a finitely generated K-algebra, each element of which is integral over K, always a finitely generated K-module?

Let us recall that an element x is called integral over K if there exist $a_1, \ldots, a_{p-1}$ in K such that

$$x^p = a_1 x^{p-1} + \cdots + a_{p-1} x$$

Or equivalently, $K[x]$ is a finitely generated K-module.

The Levitzki problem is similar: Is a finitely generated algebra, each element of which is nilpotent, always nilpotent? (That is, there exists r such that any product of r elements vanishes.) These problems admit both a negative answer (the theorem of Golod and Shafarevitch; see Chapter 8 in Herstein 1968). However, in the case of pi-algebras, the answer is affirmative; the hypothesis may even be weakened.

THEOREM 7.2.2. *Let $\mathfrak{M}$ be a finitely generated K-algebra, generated by $m_1, \ldots, m_k$ and satisfying an admissible polynomial identity of degree n. If any product of no more than $n - 1$ of the m_i's is nilpotent (resp. integral over K) then $\mathfrak{M}$ is nilpotent (resp. a finitely generated K-module).*

Proof. $\mathfrak{M}$ verifies an identity of the form shown in Lemma (7.2.1). Let $A = \{a_1, \ldots, a_k\}$ be a totally ordered alphabet and

$$\varphi: K_+\langle A \rangle \to \mathfrak{M}$$

the algebra morphism defined by $\varphi(a_i) = m_i$. Let $\mathcal{J} = \operatorname{Ker}\varphi$. Let p be an integer such that $p \geqslant 2n$ and that for any nonempty word u in A^* of length at most $n - 1$, $\varphi(u)^p = 0$ (resp. $\varphi(u)$ verifies an equation of the following type: $\varphi(u)^p = a_1\varphi(u)^{p-1} + \cdots + a_{p-1}\varphi(u)$ for some a_i in K).

Let $N = N(k, p, n)$ be the integer of Theorem 7.1.5.

Any word in A^* of length at least N contains as a factor either a pth power of some word u, $0 < |u| < n$, or an n-divided word. In the second case, by the identity displayed in Lemma 7.2.1, w is equal modulo $\mathcal{J}$ to a linear combination of words of the same length as w and which are strictly greater than w in the alphabetical order; since these words are only in finite number, we deduce inductively that w is equal modulo $\mathcal{J}$ to a linear combination of words of the same length each of which contains a pth power u^p with $0 < |u| < n$. Hence $\varphi(w) = 0$ and $\mathfrak{M}$ is nilpotent (resp. w is equal modulo $\mathcal{J}$ to a linear combination of words of length $< |w|$; hence $\mathfrak{M}$ is generated, as K-module, by the elements $\varphi(v)$, $0 < |v| < N$). ∎

7.3. Consequences

Theorem 7.2.2 is used next to prove two corollaries; their proofs use some results that are not proved in this book. By $\mathfrak{M}_n(K)$ is denoted the algebra of $n \times n$ matrices over K.

COROLLARY 7.3.1. *Let $m_1, \ldots, m_k \in \mathfrak{M}_n(K)$. If every product of no more than $2n - 1$ of the matrices m_i is a nilpotent matrix, then the semigroup generated by the m_i's is nilpotent.*

Proof. In view of Amitsur–Levitzki's theorem (see Procesi 1973, Chapter 1, Theorem 5.2) $\mathfrak{M}_n(K)$ verifies an admissible identity of degree $2n$. This identity is clearly verified by the algebra $\mathfrak{M}$ generated by the matrices m_i. Hence, by Theorem 7.2.2, $\mathfrak{M}$ is nilpotent and the corollary follows.   ∎

A semigroup S is called *periodic* if every element of S generates a finite semigroup. The *Burnside problem* (extended to semigroups) is to know whether every finitely generated periodic semigroup is finite. The answer is negative in general (even for groups, see Herstein 1968: Chapter 8). However, it was shown by Schur that the answer is positive for subgroups of $\mathfrak{M}_n(\mathbb{C})$ (see Kaplansky 1969: Theorem G), and this result has been extended to subgroups of pi-algebras (see Procesi 1973: Chapter 6, Corollary 2.8). On the other hand it has been extended to subsemigroups of $\mathfrak{M}_n(K)$ when K is a field, see McNaughton and Zalcstein 1975. It can be extended further.

COROLLARY 7.3.2. *Let K be a field and S be a finitely generated periodic subsemigroup of a pi-algebra. Then S is finite.*

Proof. For each $x \in S$, there exist integers k and p, $k \neq p$, such that $x^k = x^p$. Hence every x in S is algebraic over K. Let $\mathfrak{M}$ be the subalgebra generated by S. $\mathfrak{M}$ verifies a polynomial identity, hence by Theorem 7.2.2, $\mathfrak{M}$ is finite dimensional over K. From that we deduce that $\mathfrak{M}$ admits a faithful representation by matrices over K and the problem is reduced to the Burnside problem for subsemigroups of $\mathfrak{M}_n(K)$ (McNaughton and Zalcstein 1975).   ∎

Notes

All the results of Section 7.1 are due to Shirshov; they can be found in Shirshov 1957a except Theorem 7.1.5, which uses an argument extracted from Shirshov 1957b.

Lemma 7.2.1 is classical, see Shirshov 1957a, Chapter 6 in Herstein 1968, or Chapter 1 in Procesi 1973. Theorem 7.2.2 is from Shirshov 1957b. In the same paper he proves a more general result, which is the material of Problems 7.1.1, 7.1.2, and 7.2.1. An algebraic proof of Theorem 7.2.2. can be found in Procesi 1973, and the same proof as here (Shirshov's original proof) may also be found in Rowen 1980, § 4.2. For further information about the Kurosch and Burnside problems see Kaplansky 1969; see also Chapter 2.

About Corollary 7.3.1: If K is a field and S a subsemigroup of $\mathfrak{M}_n(K)$ each matrix of which is nilpotent, S can be put into triangular form as shown by Levitzki; see Theorem 32 in Kaplansky 1969, which implies that the nilpotency index of S is $< n$; that is, $S^n = 0$. Furthermore the corollary shows that if S is not nilpotent then there exists a nonnilpotent matrix in S, equal to a product of no more than $2n-1$ of the generators. A related result is a theorem of Jacob (1980): there exists an integer N (depending on K, n, and the number of generators) and a pseudo-regular matrix in S, equal to a product of no more than N of the generators; see Problem 7.3.1. It has been extended in Reutenauer 1980: From any long product of any matrices one can extract a subproduct that is a pseudo-regular matrix.

Problems

Section 7.1

7.1.1. Let $u_0, \ldots u_n, v, w$ be words, with v nonempty and w not a left factor of v. Show that the word

$$u_0 v^n w u_1 v^n w \cdots u_{n-1} v^n w u_n$$

admits an n-divided factor.

7.1.2. A language L (that is a subset of A^+) is *bounded* if there exist words $f_1, \ldots, f_k$ such that

$$L \subset f_1^* f_2^* \cdots f_k^*.$$

Notice that each finite union of bounded languages is bounded. Show that A^* is not bounded (use Chapter 2). Let $n \geqslant 1$. Show that the set of words admitting no n-divided factor is a bounded language (use the previous problem and Theorem 7.1.5).

7.1.3. A *quasi-power of order* 0 is any non-empty word. A *quasi-power of order* $n+1$ is a word of the form uvu where u is a quasi-power of order n. Show that given a (finite) alphabet A, there exists a sequence of integers $N(n)$ such that each word on A of length at least $N(n)$ contains a factor that is a quasi-power of order n.

Section 7.2

7.2.1. Let $\mathfrak{M}$ be a K-algebra and

$$\varphi: K_+ \langle A \rangle \to \mathfrak{M}$$

a surjective morphism (K is a commutative ring and A a finite alphabet). Suppose that $\mathfrak{M}$ is a pi-algebra. Show that there is a bounded language L such that $\mathfrak{M}$ is equal to the submodule generated by $\varphi(L)$ (use the previous problem). Deduce Theorem 7.2.2. again. (Reference Shirshov 1957: Theorem 1.)

7.2.2. Let K be a commutative ring with field of fractions F. Let S be a finitely generated multiplicative subsemigroup of $\mathfrak{M}_n(F)$ such that each $m \in S$ is integral over K. Show that there exists d in $K, d \neq 0$, verifying: $dS \subset \mathfrak{M}_n(K)$.

Section 7.3

7.3.1. A matrix m is *pseudo-regular* if it is contained in a subgroup of the multiplicative semigroup of $\mathfrak{M}_n(K)$, where K is a field. Show that it is equivalent to: $Ker\, m \cap Im\, m = 0$ (letting m act on the right on K^n). Show that if u, v are matrices such that $\mathrm{rank}(u) = \mathrm{rank}(uvu)$ then vu is pseudo-regular. Let

$$\mu: A^* \to \mathfrak{M}_n(K)$$

be a monoid morphism from a finitely generated free monoid into the multiplicative monoid of n by n matrices over K. Show that if w is a word of length at least $N(n)$ (see Problem 7.1.3) then w contains a factor $u \neq 1$ such that μu is pseudo-regular.

CHAPTER 8

The Critical Factorization Theorem

8.1. Preliminaries

Periodicity is an important property of words that is often used in applications of combinatorics on words. The main results concerning it are the theorem of Fine and Wilf already given in Chapter 1 and the Critical Factorization theorem that is the object of this chapter. Both admit generalization to functions of real variables, a topic that will not be touched here.

In all that follows, we shall be considering a fixed word $w = a_1 a_2 \cdots a_n$ of positive length n, where $a_1, a_2, \ldots, a_n \in A$. The *period* $p = \pi(w)$ of w is the minimum length of the words admitting w as a factor of some of their powers; these words of minimal length p are the *cyclic roots* of w. Furthermore, w is *primitive* if it is not a *proper* power of one of its cyclic roots, *primary* iff it is a cyclic root of itself, and *periodic* iff $n \geq 2p$.

For instance $w = aaabaa$ of length 6 has period 4; its cyclic roots are *aaab, aaba, abaa, baaa*. It is not primary or periodic; its cyclic roots are all primitive, but only *aaab* and *baaa* are primary. The word *aaabaaabaa* is periodic; it is also primitive, and its cyclic roots are the same as for w.

It is clear that the set (denoted $\sqrt{w}$) of the cyclic roots of w is a conjugacy class including all the factors $a_i a_{i+1} \cdots a_{i+p-1}$ of length p of w. It contains p distinct primitive words and, accordingly, some of them are not factors of w when w is not periodic. We leave it to the reader to check the following statement:

PROPOSITION 8.1.1. *Equivalent definitions of the period p of w are*

  (i) $p = n - |v|$, *where v is the longest word $\neq w$ that is a left and a right factor of w;*

 (ii) $p = n$ *iff w is primary; otherwise it is the least positive integer such that one has identically $a_i = a_{i+p}$. ($1 \leq i < i + p \leq n$.)*

For the sake of completeness the theorem of Fine and Wilf will be reformulated in the present notation. An algebraic proof was given in Chapter 1.

THEOREM 8.1.2. *Let* $w = w'uw''$ *(with* $w', w'' \in A^*$, $u \in A^*$), $p' = \pi(w'u)$, $p'' = \pi(uw'')$, *and d be the greatest common divisor of* p' *and* p''.

$$If \; |u| \geq p' + p'' - d \; one \; has \; p' = p'' = \pi(w).$$

8.2. The Critical Factorization Theorem

Let us consider a given factorization of the fixed word $w = a_1 \cdots a_n$ into two factors $w' = a_1 \cdots a_i$ and $w'' = a_{i+1} \cdots a_n$. We always assume that it is *proper*; that is, $w', w'' \neq 1$.

The set of the *cross factors* of (w', w'') is by definition

$$C(w', w'') = \{ u \in A^+ \mid A^*u \cap A^*w' \neq \emptyset \quad and \quad uA^* \cap w''A^* \neq \emptyset \}$$

The minimum of the length of the cross factors of (w', w'') is the *virtual period* $p(w', w'')$ of the factorization.

The factorization (w', w'') is *critical* iff $p(w', w'')$ is equal to the period p of w.

An explanation of this formal definition is needed. Assume for instance $i \geq p$, the period of w. Then the cyclic root $u = a_{i-p+1} \cdots a_{i-1}a_i$ is a *right* factor of the *left* term $w' = a_1 \cdots a_i$ of the factorization; that is, $w' \in A^*u$, implying $A^*u \cap A^*w' \neq \emptyset$. Let us say that it is *left internal*. When $|w''| \geq p$ the same situation is obtained symmetrically on the right (that is, for the right term $w'' = a_{i+1} \cdots a_n$) with the same word u when $|w''| \geq p$, because $a_{i-p+1} = a_{i+1}$, $a_{i-p+2} = a_{i+2}, \ldots, a_i = a_{i+p-1}$. When $|w''| < p$, the same equations show that w'' is equal to a left factor of u. Thus we have $u \in w''A^*$, implying again $uA^* \cap w''A^* \neq \emptyset$. In this case we say that u is *right external*.

The same applies with obvious modifications when $i < p$. We see that the set $C = C(w', w'')$ of the cross factors contains always at least one cyclic root of w, hence that the virtual period of a factorization never exceeds the true period of the word.

In fact it is "usually" far smaller. For example if $w = a^n b^m$ (with $n, m \geq 1$), the virtual period of any factorization is 1 (and it is internal on both sides) except for the factorization $(w' = a^n, w'' = b^m)$ where it attains its maximal value $n + m = \pi(w)$ (and it is external on both sides).

A more typical example is that of the word $w = aacabaca$ $(n = 8, p = 7)$. The successive virtual periods ρ_i of the factorizations (w_i', w_i'') $((|w_i'| = i, |w_i''| = n - i) \; 1 \leq i \leq 7)$ are indicated below:

$$w = a_1 a_7 c_7 a_7 b_4 a_4 c_2 a$$

For instance, if $i = 5$ (that is, $w_5' = aacab, w_5'' = aca$), one finds $\rho_5 = 4$, because this is the length of the shortest corresponding cross factor $u_5 = acab$ (which is left internal and right external). For $i = 3$, the shortest cross factor is $u_3 = abacaac$, *which* has length 7 (and is external on both sides). By our definition the factorizations for $i = 2, 3, 4$ are critical. The main theorem of this chapter can now be stated. To give it in its most useful form let us exclude the case where the period p of w is 1. Indeed, in that case, w consists of the same letter repeated n times and every factorization is critical.

THEOREM 8.2.1 (Critical Factorization Theorem). *Let w be a word of period $p > 1$. Any sequence of $p - 1$ successive factorizations $\{(w_i', w_i'') \mid j < |w_i| < j + p\}$ contains at least one critical factorization. The corresponding shortest cross factor is a primary cyclic root of w.*

A corollary is the fact that any conjugacy class of primitive words contains primary words. This fact can be proved directly considering Lyndon words (Chapter 5). The same technique allows an easy proof of the theorem in the special case $n \geq 3p$ (see Problem 5.1.2).

The theorem as given is sharp. Indeed the word $w = a^m b a^m$ $(m \geq 1)$ has period $p = m + 1$ and exactly two critical factorizations, viz. (a^m, ba^m) and $(a^m b, a^m)$. Only one of them is contained in the sequence of the $p - 1$ first (or last) factorizations.

The proof is by induction on the length of w. It is more or less an existence proof because we do not know how to find the critical factorizations (or the primary words) other than by sifting out the other ones. Thus we proceed leisurely in order to provide the maximum information on the structure of the word.

We keep the same notations and let $C = C(w', w'')$.

PROPOSITION 8.2.2. *A cross factor $u \in C$ of minimal length is primary.*

Proof. Suppose that $u \in C$ is not primary—that is, that $u = u'g'' = g'u'$ for some $u', g', g'' \in A^*$. One has $u \in A^*u'$, hence $A^*u \subset A^*u'$ implying $A^*u' \cap A^*w' \neq \emptyset$. The same holds on the right showing that u' is also a cross factor. Since $|u'| < |u|$ we conclude that u has not minimal length. ∎

PROPOSITION 8.2.3. *The set C contains a unique shortest cross factor u. If $|u| = p$, u is a cyclic root of w. This condition is satisfied if u is external on both sides.*

Proof. Note that any right (left) internal cross factor in C is strictly shorter than any right (left) external one. Thus a cross factor u of minimal length is internal on at least one side except if every cross factor in C is external on both sides—that is except if $C \subset A^*w' \cap w''A^*$.

Since internal cross factors are true factors of w, this proves the unicity of the minimal $u \in C$ and the fact that it is a cyclic root of w except in the completely external case which we consider now.

Let $v \in A^*$ (that is, possibly $v = 1$) be the longest word that is a *right* factor of the *right* term w'' ($w'' \in A^*v$) and a *left* factor of the left term w' ($w' \in vA^*$). We have $w' = vy'$, $w'' = y''v$, where y', $y'' \in A^+$, because otherwise w' or w'' would itself be a cross factor, contradicting the hypothesis that every cross factor is external.

Set $u = y''vy' = y''w' = w''y'$. By construction $u \in C$ and by our choice of v it is the (unique!) shortest word in $A^+w' \cap w''A^+$. Since C is contained in this last set by hypothesis, u is indeed the desired shortest cross factor.

We verify that u is a cyclic root of w. Observe first that no word of length $k \geq |w'|$ or $k \geq |w''|$ is a left and a right factor of w, because otherwise we would have $w' \in C$ or $w'' \in C$. In view of y', $y'' \neq 1$, this shows that v is the longest left and right factor of w—that is, that $w = vy'y''v$ has period $|vy'y''| = |u|$.   ∎

The next result is the key to the proof.

LEMMA 8.2.4. *Let $y \in A^*$ be a right (left) factor of the left (right) term w' (w'') of the factorization and let $q = \pi(y)$ be its period. If u is the shortest cross factor of (w', w'') one has $|u| \leq q$ or $|y| < |u|$.*

Proof. Since u must be primary it suffices to take a cross factor u' satisfying the opposite inequalities $q < |u'| \leq |y|$ and to verify that u' is not primary.

In view of $u' \in C$, $w' \in A^*y$, and $|u'| \leq |y|$, we have $y = y'u'$. Because of $q < |u'|$ we can write $u' = u''z$ with $|z| = q$, $u'' \in A^+$. However, since q is the period of y its left factor $y'u''$ is also a right factor—that is, $u' \in A^+u''$—showing that u' is not primary.   ∎

PROPOSITION 8.2.5. *Let $b \in A$ be a letter and assume that (w', w'') is a critical factorization of w.*

If w and wb have the same period, $(w', w''b)$ is a critical factorization of wb. In the opposite case, $(w', w''b)$ is critical iff (w', w'') is right external and, then (w', w'') is external on both sides.

Proof. Let u as before and v be the shortest cross factor of $(w', w''b)$.

In view of the minimum character of u and v one has $|u| \leq |v|$ with equality iff $u = v$.

Suppose first that u is right internal; that is, $w'' \in uA^*$. It is also a cross factor of $(w', w''b)$; hence $v = u$ and the factorization $(w', w''b)$ of wb is critical iff $\pi(wb) = p$.

Suppose on the contrary that $u = w''cx$ for some letter $c \in A$ and word $x \in A^*$. Because of $|v| \geq |u|$, v is not a left factor of w'' and accordingly it has the form $v = w''bx'$ for some $x' \in A^*$. The hypothesis that (w', w'') is critical is equivalent with $u \in (\sqrt{w})$ (the set of cyclic roots of w), that is with $c = a_{n-p+1}$. Thus when wb has the same period p as w —that is, when $b = a_{n-p+1}$—we have $b = c$; hence, as in the first case, $v = u$ and $(w', w''b)$ is critical.

Therefore we can assume $\pi(wb) \neq p, c \neq b$ and consequently $|v| > |u|$. Note first that v is right external because $|u| > |w''|$ and $|w''b| = |w''| + 1$.

To complete the proof it suffices by Proposition 8.2.3 to show that v is left external; that is, that $|v| > |w'|$. However, this follows from Lemma 8.2.4. with w' instead of y and $w''b$ and v instead of w'' and u, since we have $|v| > |u| = \pi(w) \geq \pi(w')$.   ∎

COROLLARY 8.2.6. *Under the assumption that* $\pi(a_2 \cdots a_n b) \leq p$, *every critical factorization* (w', w'') *of* $w = a_1 \cdots a_n$ *gives a critical factorization* $(w', w''b)$ *of* wb.

Proof. Let $p' = \pi(wb)$, $\overline{w} = a_2 \cdots a_n$, and $\overline{p} = \pi(\overline{w}b)$. By the last result it suffices to show that when $p \neq p'$ the hypothesis $\overline{p} \leq p$ implies that every critical factorization of w is right external. This condition is certainly satisfied when $p = n$ since then the shortest cross factor has length $n = |w|$. Thus we have only to consider the case when $\overline{p} \leq p < n$ and $p < p'$.

We cannot have $\overline{p} = p$ because this would entail $b = a_{n-1+p}$; that is, $p = p'$. Thus $\overline{p} < p$. Consider any critical factorization (w', w'') of w. The word w'' is a factor of $\overline{w}$, hence of $\overline{w}b$ and, consequently its period q is at most $\overline{p} < p$. Applying Lemma 8.2.4 with $y = w''$, we conclude that the length $|u|$ of the shortest cross factor satisfies $|u| > |w''|$—that is, that this factorization is right external.   ∎

Proof of Theorem 8.2.1. The theorem is easily verified for words of short length. Thus, in order to keep the same notation, we shall assume that the theorem is already established for words of length n, $n \geq 2$, and, letting $w = a_1 \cdots a_n$ as before, it will suffice to verify its truth for wb ($b \in A$). We set $\overline{w} = a_2 \cdots a_n$, $\overline{p} = \pi(\overline{w}b)$, $p^* = \pi(wb)$.

By symmetry we can suppose $\overline{p} \leq p$ ($= \pi(w)$) and we distinguish three cases:

(i) $\overline{p} = p = n < p^*$ ($= n + 1$)
(ii) $\overline{p} = p = p^*$
(iii) $\overline{p} < p$

There is no further case. Indeed under the assumption $\overline{p} = p < n$, one has $b = a_{n-p+1}$ (since $n - p + 1 \geq 2$) and $a_1 = a_{1+p}$ (since $1 + p \leq n$) implying that wb has the same period p as w and $\overline{w}b$.

In each of the three cases, Proposition 8.2.5 and Corollary 8.2.6 show that every critical factorization (w', w'') of w gives a critical factorization $(w', w''b)$

of wb. It remains to verify that any sequence of $p^* - 1$ factorizations of wb contains a critical one. Since $p^* \geqslant p$ this follows instantly from the induction hypothesis on w provided that $p^* > p$, or more generally, provided that one has ascertained the existence of a critical factorization of wb that is right external.

Case 1. By the induction hypothesis w has a critical factorization. It is right external since $p = n$. Thus it gives a critical factorization of wb, and the full result follows since $\pi(wb) = n + 1$.

Case 2. Since $\pi(w) = \pi(wb)$ every critical factorization of w gives one of wb. By symmetry the same holds for $\overline{w}b$. Thus by the induction hypothesis on $\overline{w}b$ we conclude that wb has a critical factorization $(w', w''b)$ with $|w''b| < p$, and the result is entirely proved. An example of this case is

$$w = a^{p-1}ba^{p-1}, \ (p \geqslant 2).$$

Case 3. Consider a critical factorization (w', w'') of w. Since $w' \neq 1$, w'' is a factor of $\overline{w}$, hence of $\overline{w}b$ and accordingly its period q is at most $\overline{p} < p$. Applying Lemma 8.2.4 with $y = w''$, we see that the shortest cross factor u must satisfy $|u| > |w''|$. Thus it is right external and, accordingly, by Proposition 8.2.5, $(w', w''b)$ is a critical factorization of wb. Now the same lemma with $y = w''b$ shows that the corresponding cross factor is strictly longer than $w''b$, proving that $(w', w''b)$ is also external and concluding the proof.   ∎

Consider the set J of the factorizations of w that are not internal on both sides. They belong to three types: left external and right internal, left and right external, and left internal and right external.

The next property shows that the factorizations appear in this order when w is read from left to right. Also, the sequence $(p_j : j \in J)$ of the virtual periods corresponding to the factorizations in J is unimodal in the sense that if $j < j' < j''$ one cannot have $p_j > p_{j'} < p_{j''}$. For simplicity the property is stated on one side only.

PROPOSITION 8.2.7. *Let u and v be the shortest cross factors of the factorizations (w', xw''') and $(w'x, w''')$ $(x \in A^*)$ of w and assume that u is left internal and right external. Then v is left internal. Further, $|u| \geqslant |v|$ with equality only if v is right external and a conjugate of u.*

Proof. Let $w'' = xw'''$. The hypothesis means that there is a word $y \in A^*$ such that $w''y$ is a right factor of the left term w' and that no strictly shorter right factor of w' is a left factor of a word in $w''A^*$.

Thus the left term $w'x$ of the factorization $(w'x, w''')$ has at least one right factor, viz. $w'''yx$, which belongs to $w'''A^*$. This shows that the minimal cross factor v is left internal. We have the inequalities

$$|v| \leqslant |w''yx| = |xw''y| = |u|,$$

which prove that the virtual period $|v|$ is at most equal to the virtual period $|u|$ with equality iff $v = w'''yx$. Since $w'''yx$ is a conjugate of $u = xw'''y$, the equality $|v| = |u|$ entails that v be right external as u. ∎

8.3.   An Application

This section gives an application of the critical factorization theorem to a problem arising in the study of free submonoids and free groups.

Let $w = a_1 \cdots a_n$ be a fixed word, as before, and let X be a given finite set of words of positive length. In the case of interest the words of X are short with respect to w and we assume $|x| \leqslant n$ for all $x \in X$.

The problem is to find upper bounds to the number of different ways w can appear as factor in a word of the submonoid X^* generated by X or, in equivalent manner, to bound the number of factorizations of w of the form $w = x'yx''$ with $y \in X^*$ and x'' (resp. x') a *prefix* (resp. *suffix*) of X—that is, a proper *left* (resp. *right*) factor of a word of X that is not itself a word of X.

For instance, let $w = bababbaba$, $X = \{a, bb, bab\}$. Two such factorizations can be found: $w = [(bab)(a)(bb)(a)]ba$ and $w = b[(a)(bab)(bab)(a)]$, where the term y is between square brackets and the pair x'', x' is respectively $(1, ba)$ and $(b, 1)$.

A *X-interpretation* of w is a factorization $w = x_0 x_1 \cdots x_r x_{r+1}$ where $0 \leqslant r$, $x_i \in X$ for $1 \leqslant i \leqslant r$, x_{r+1} (resp. x_0) is a prefix (resp. suffix of X). It is *disjoint* from another X-interpretation $y_0 y_1 \cdots y_s y_{s+1}$ iff $x_0 x_1 \cdots x_i \neq y_0 y_1 \cdots y_j$ for any $i \leqslant r$, $j \leqslant s$. The *X-degree* of w is the maximum number of elements in a system of pairwise disjoint X-interpretations of w.

The main result of this section may be stated as follows:

THEOREM 8.3.1. *If the period of w is strictly greater than the periods of the words of X, the X-degree of w is at most Card(X).*

The restriction upon the periodicity of w is clearly necessary to obtain such a bound. The word $w = a^n$ has $p + 1$ X-interpretation when X consists of a^p only. The notion of disjointedness has algebraic motivations that have no place here. It suffices to note that without this restriction the number of X-interpretations of w could grow exponentially with its length when there are words that admit several factorizations as product of words of X (that is, when X is not a code).

The bound given by the theorem is not far from being sharp, and it is even conjectured that the exact value is $-1 + \text{Card}(X)$. Indeed if X is the biprefix code made up of a^p and of p words of the form $a^i b a^j$ where i and j

take each of the values $0, 1, \ldots, p-1$ exactly once, it is easily verified that any word w admits exactly p pairwise disjoint X-factorizations provided that two successive bs are separated by more than $2p-2$ a's (that is, $w \in (a^*a^{2p-2}b)^*a^*$).

The proof of the theorem uses a lemma of independent interest. To handle conveniently the various occurrences of the same word as a factor of w, we need one more definition: a *covering*.

Let $v = a_i \cdots a_j$ ($i \leqslant j$) be a segment of w and x be a word of length at least $j+1-i$. A *covering* of v by x is a segment $v' = a_{i'} \cdots a_{j'}$ such that $i' \leqslant i \leqslant j \leqslant j'$ and that $v' = x$ or that $i' = 1$ (resp. $j' = n$) with v' a proper right (resp. left) factor of x.

Let now $w = w'vw''$ where $v = a_i \cdots a_j$ as above and where $w', w'' \in A^*$.

LEMMA 8.3.2. *If the segment v has two coverings by x, the virtual period of any factorization $(w'v', v''w'')$ $(v', v'' \in A^*, v'v'' = v)$ is at most equal to the period of x.*

Proof. Let $a_{i_1} \cdots a_{j_1}, a_{i_2} \cdots a_{j_2}$ be the two coverings of v by x. We can suppose $i_1 \leqslant i_2$ and $j_1 \leqslant j_2$, where at least one inequality is strict.

There are several cases to consider. Assume first that both words $a_{i_1} \cdots a_{j_1}$ and $a_{i_2} \cdots a_{j_2}$ are equal to x. Thus $i_2 - i_1 = j_2 - j_1$ and the common value d of this difference is strictly less than $|x|$ because the two intervals $[i_1, j_1]$ and $[i_2, j_2]$ overlap on $[i, j]$ or on a larger interval. Also we have $a_k = a_{k+d}$ for every k satisfying $i_1 \leqslant k < k + d \leqslant j_2$. It follows that for any factorization $(w'v', v''w'')$ $(v'v'' = v)$, the right factor of length d of $w'v'$ is equal to the left factor of the same length of $v''w''$.

It also follows that the period p of x is at most d. Suppose $p < d$ and let q be the length of the shortest cross factor of $(w'v', v''w'')$. By Lemma 8.2.4 we have $q \leqslant p$ or $a \geqslant |w'v'|$. Since $q \leqslant d$ where d is at most equal to the length of $w'v'$ we conclude that $q \leqslant p$ and the result is proved in this case.

Assume now that $a_{i_1} \cdots a_{i_2} = x''$, a proper right factor of x, but that the same is not true for $a_{i_2} \cdots a_{j_2}$. Defining x' by $x'x'' = x$ and replacing w' by $x'w'$ cannot decrease the virtual period. The same can be done for w'' if the covering $a_{i_2} \cdots a_{j_2}$ is right external, and we are back to the initial case.

It remains only to discuss the case where both $a_{i_1} \cdots a_{j_1}$ and $a_{i_2} \cdots a_{j_2}$ are external on the same side, say on the left. Then we have $j_1 < j_2$ and $a_1 \cdots a_{j_1} = x_1'$ equal to a right factor of the right factor $a_1 \cdots a_{j_2} = x_2'$ of x. It is clear that in fact x enters in the discussion only through x_2'; replacing it by x_2' we are back to the previous case, where only one of the two coverings is not internal. The conclusion is that the virtual period is bounded by the period of x_2', which is itself at most equal to that of x, and the result is proved in all cases. ∎

Proof of Theorem 8.3.1. Let (w', w'') be a critical factorization of w. We have $1 \leqslant |w'| = i \leqslant |w| - 1$.

For each X-interpretation $w = x_0 x_1 \cdots x_{r+1}$ $(0 \leqslant r)$, there is a smallest s such that $|x_0 x_1 \cdots x_s| \geqslant i$ and a word $x = x_j$ that produces a covering of the segment $v = a_i$ of length 1 of w. If there exists a system of $1 + \operatorname{Card} X$ pairwise disjoint X-interpretations, at least two of them lead to the same word x and the conclusion follows from the preceding lemma. ∎

Notes

The critical factorization theorem and its application in Section 8.3 have been discovered by Cesari and Vincent (1978). The presentation given here follows closely Duval (1979), who contributed many crucial improvements and developed a deeper theory in his thesis (1980). Further applications also due to Duval relate the period of a word to the maximum length of its primary factors, a problem originally attacked in Ehrenfeucht and Silberger 1979 and in Assous and Pouzet 1979. See Duval 1980 for the best result known so far in this direction.

CHAPTER 9

Equations in Words

9.0 Introduction

Let us consider two words x, y of the free monoid A^*, satisfying the equality:

$$xy = yx. \tag{9.0.1}$$

By Proposition 1.3.2 of Chapter 1, there exist a word $u \in A^*$ and two integers $n, p \geq 0$ such that

$$x = u^n \quad \text{and} \quad y = u^p \tag{9.0.2}$$

In this chapter, we will view x and y as the letters of an alphabet Ξ. We will say that $xy = yx$ is an equation in the unknowns $\Xi = \{x, y\}$ and that the morphism α: $\Xi^* \to A^*$ defined by $\alpha(x) = u^n$ and $\alpha(y) = u^p$ is a solution of the equation. Observe that all solutions of this particular equation are of this type.

The basic notions on equations are presented in Section 9.1. In Section 9.2, we consider a few equations whose families of solutions admit a finite description, as in the preceding example. Indeed, the family of solutions of Eq. (9.0.1) is entirely described by the unique expression (9.0.2), where u runs over all words and n, p over all positive integers. This idea is formalized in Section 9.3, which introduces the notion of parametrizable equations and where it is recalled that all equations in three unknowns are parametrizable.

Not all equations are parametrizable, however. We are thus led in Section 9.4 to define the rank of an equation, which is the maximum number of the letters occurring in the expression of particular solutions called *principal*. An effective construction of these solutions is given in Section 9.5 that in Section 9.6 leads to the notion of graph associated with an equation. This allows in Section 9.7 the effective calculation of the rank of all quadratic equations—that is, the calculation of all equations for which each letter has exactly one occurrence in each member of the equation. Finally, in Section 9.8 it is shown how the present theory is related to some other ones, such as theories of equations with constants, of test sets, and so on.

9.1. Preliminaries

Consider a fixed, finite, nonempty alphabet $\Xi = \{x_1, \ldots, x_p\}$, and a set S of pairs of words $(e, e') \in \Xi^* \times \Xi^*$. Let us ask for all morphisms α of Ξ^* into some free monoid A^* such that $\alpha(e) = \alpha(e')$ holds for all $(e, e') \in \mathsf{S}$. It is natural to speak of S as a system of *equations in Ξ^** (or equivalently *in the unknowns $x_1, \ldots, x_p$*) and of α as a *solution* of this system, over the free monoid A^*. Let us also say that α *satisfies* S.

Very little can be found in the literature on infinite systems of equations, apart from some results in the theory of test sets, which will be presented in the last section of this chapter. For *finite* systems of equations, there is no loss of generality in restricting consideration to single equations. Indeed, let us say that a morphism $\alpha \colon \Xi^* \to A^*$ is *cyclic* if there exists $v \in A^*$ such that $\alpha(x) \in v^*$ holds for all $x \in \Xi$. Then the following can be established (see Hmelevskii 1976):

PROPOSITION 9.1.1. *Let S be a finite system of equations in Ξ^*. Then there exists an equation $(e, e') \in \Xi^* \times \Xi^*$ such that for all noncyclic morphisms $\alpha \colon \Xi^* \to A^*$, we have: α satisfies S iff it satisfies (e, e').*
It is worth observing that in the proof of the proposition, Hmelevskii effectively constructs (e, e') from S.

The problem of solving equations in free monoids was preceded by the corresponding problem in free groups. One of the oldest problems posed was indeed as follows: What are the morphisms α of the free group generated by the three elements x, y, z, into an arbitrary free group F, for which $\alpha(x^2 y^2 z^2) = 1$ holds? The theory of equations has since been extended to various classes of monoids (for example, by Putcha 1979). It should be remembered that, because of the natural embedding of Ξ^* in the free group G generated by the elements of Ξ, all solutions of an equation in Ξ^* can be viewed as a solution of the corresponding equation in G. An application of this remark can be seen in Section 9.2.

From now on we shall consider only single equations (e, e'). Further, unless otherwise mentioned, we shall assume that each element in Ξ has an occurrence in either e or e'. In the same way, if $\alpha \colon \Xi^* \to A^*$ is a solution, we shall assume that each element in A occurs either in $\alpha(e)$ or in $\alpha(e')$. Thus A is finite.

Since solutions of an equation are defined as morphisms satisfying some fixed relation, we need some elementary notions involving morphisms.

Given a morphism $\alpha: \Xi^* \to A^*$, we say that it is

- *Total*, if each letter of A has some occurrence in a word $\alpha(x)$, for some $x \in \Xi$;
- *Nonerasing* (see Chapter 1), if $\alpha(x) \neq 1$ holds for all $x \in \Xi$;
- *Cyclic*, if there exists a word $v \in A^*$ such that $\alpha(x) \in v^*$ holds for each $x \in \Xi$;
- *Trivial*, if $\alpha(x) = 1$ holds for all $x \in \Xi$.

Notice that a total morphism $\alpha: \Xi^* \to A^*$ is trivial iff $A = \varnothing$.

Given two total morphisms, $\alpha_1: \Xi^* \to A_1^*$ and $\alpha_2: \Xi^* \to A_2^*$, we say that α_1 *divides* α_2 and we write $\alpha_1 \leq \alpha_2$ iff there exists a nonerasing morphism $\theta: A_1^* \to A_2^*$ satisfying $\alpha_2 = \theta \circ \alpha_1$. We say that α_1 and α_2 are *equivalent*, and we write $\alpha_1 \approx \alpha_2$, iff $\alpha_1 \leq \alpha_2$ and $\alpha_2 \leq \alpha_1$. In fact it can easily be seen that α_1 and α_2 are equivalent iff they are equal up to a renaming of the alphabets A_1 and A_2 — that is, iff there exists a morphism $\theta: A_1^* \to A_2^*$, mapping bijectively A_1 onto A_2, satisfying $\alpha_2 = \theta \circ \alpha_1$ and thus $\alpha_1 = \theta^{-1} \circ \alpha_2$. In the sequel, unless otherwise stated, we shall not distinguish between two equivalent morphisms.

The *rank* of a morphism $\alpha: \Xi^* \to A^*$, denoted rank α, is the cardinality of the basis of the free hull of the set $\alpha(\Xi) = \{\alpha(x) \in A^* \mid x \in \Xi\}$ (see Chapter 1). Obviously, two equivalent morphisms have the same rank.

PROPOSITION 9.1.2. *Given two arbitrary morphisms* $\alpha: \Xi^* \to A^*$ *and* $\theta: A^* \to B^*$, *we have*

$$\text{rank}(\theta \circ \alpha) \leq \text{rank}(\alpha).$$

Proof. Let $X = \alpha(\Xi)$ and $Y = \theta(X)$. Let U, V be the free hulls of X and Y respectively. Let W be the free hull of $\theta(U)$. Then $X \subset U^*$ implies $Y \subset W^*$. Since W^* is free, we have $V^* \subset W^*$ by the definition of the free hull. Further, since $Y \subset V^*$, we have $X \subset \theta^{-1}(V^*)$. Since V^* is free, $\theta^{-1}(V^*)$ is free (see Problem 1.2.4). Therefore we have $U \subset \theta^{-1}(V^*)$. This implies $\theta(U) \subset V^*$ and consequently $W^* \subset V^*$. We have thus proved that $V = W$. By the definition, rank $(\theta \circ \alpha) = \text{Card}(V)$ and rank $(\alpha) = \text{Card}(U)$. We have, by the defect theorem (1.2.5), $\text{Card}(W) \leq \text{Card}(\theta(U))$. Since $V = W$ and $\text{Card}(\theta(U)) \leq \text{Card}(U)$, we obtain $\text{Card}(V) \leq \text{Card}(U)$, which was to be proved.   ■

9.2. A Classical Equation: $(x^n y^m, z^p)$

As a direct consequence of the defect theorem (Theorem 1.2.5) equations in two unknowns have only cyclic solutions. Solving these equations thus amounts to determining all solutions of some linear homogeneous diophantine equation. It is not hard to see that all such solutions are linear combinations of finitely many "minimal" solutions that can be constructed effectively (see Problem 9.4.4).

For equations with more than two unknowns, the situation is much more complex, as the next sections show. Solved here are a few equations in three and four unknowns that admit noncyclic solutions. We will concentrate on the equation $(x^n y^m, z^p)$ and show that, provided n, m, p are greater than 1, it admits only cyclic solutions (for the case $n = 1$ or $m = 1$ see Problem 9.2.1).

This last result plays a crucial role in studying the morphisms θ of $\{a, b\}^*$ into itself, for which $\theta(u)$ is imprimitive only if u is itself imprimitive. Lentin and Schützenberger (1967) proved that these morphisms are characterized by the fact that for all words u in the set $a^* b \cup b a^*$, $\theta(u)$ is primitive.

Appel and Djorup (1968) have solved a related equation. Indeed consider $\Xi = \{x_1, \ldots, x_k, y\}$ and let n be an integer greater than or equal to k. Those authors show that the equation $(x_1^n \ldots x_k^n, y^n)$ admits only cyclic solutions.

Lyndon and Schützenberger (1962) have proved that the equation $(x^n y^m, z^p)$ in free groups, admits only cyclic solutions, provided $n, m, p \geqslant 2$. In particular this implies the same result for free monoids. The case of free monoids will be treated after the solving of a few particular equations that will help in establishing the result.

PROPOSITION 9.2.1. *For all solutions* $\alpha: \Xi^* \to A^*$ *of the equation* (xyz, zxy), *there exist two words* $u, v \in A^*$ *and integers* $i, j, k \geqslant 0$ *satisfying*:

$$\alpha(x) = (uv)^i u, \qquad \alpha(y) = v(uv)^j \quad \text{and} \quad \alpha(z) = (uv)^k.$$

Proof. Consider the new alphabet $\Theta = \{a, b\}$ and denote by $\varphi: \Theta^* \to \Xi^*$ the morphism defined by

$$\varphi(a) = xy \quad \text{and} \quad \varphi(b) = z$$

For any solution $\alpha: \Xi^* \to A^*$ of the equation (xyz, zxy), the morphism $\alpha \circ \varphi: \Theta^* \to A^*$ is a solution of the equation (ab, ba). By the defect theorem, this means that there exist $u \in A^*$ and integers $n, m \geqslant 0$ such that $\alpha[\varphi(a)] = u^n$ and $\alpha[\varphi(b)] = u^m$.

Equality $\alpha(xy) = \alpha(x)\alpha(y) = u^n$ implies that for some $u_1, u_2 \in A^*$ and some integer $0 \leqslant p < n$ we have $\alpha(x) = (u_1 u_2)^p u_1$, $\alpha(y) = u_2 (u_1 u_2)^{n-p-1}$, and $u = u_2 u_1$, which yields the result. ∎

PROPOSITION 9.2.2. *For all solutions* $\alpha: \Xi^* \to A^*$ *of the equation* $(xy^2 x, zt^2 z)$, *there exist two words* $u, v \in A^*$ *and integers* $i, j, k, l \geqslant 0$ *such that*

$$\alpha(x) = (uv)^i u, \qquad \alpha(y) = v(uv)^j, \qquad \alpha(z) = (uv)^k u,$$

$$\alpha(t) = v(uv)^l, \quad \text{and} \quad i + j = k + l.$$

Proof. Set

$$\alpha(x) = a, \qquad \alpha(y) = b, \qquad \alpha(z) = c, \qquad \alpha(t) = d.$$

Then we obtain $ab^2a = cd^2c$. Arguing on the lengths of these words we conclude that this equality splits in two: $ab = cd$ and $ba = dc$. Without loss of generality we may assume that $|a| \geq |c|$. Then there exists $e \in A^*$ with $a = ce$ and $d = eb$. After substitution in $ba = dc$, we get $bce = ebc$. Because of the preceding proposition, there exist $u, v \in A^*$ and $i, j, k \geq 0$ such that $b = (uv)^i u$, $c = v(uv)^j$, and $e = (uv)^k$. This completes the proof. ∎

Consider now the equation $((xy)^m x, z^n)$ where for obvious reasons we assume $m > 0$ and $n > 1$. Solving it amounts to determining under what conditions a periodic word, as defined in Chapter 8, is imprimitive.

Assume now $m > 1$ and let $\alpha: \Xi^* \to A^*$ be a solution. Then by Proposition 1.3.5 of Chapter 1, there exist a word $u \in A^*$ and two integers $i, j \geq 0$ satisfying $\alpha(xy) = u^i$ and $\alpha(z) = u^j$. This yields

$$\alpha((xy)^m x) = u^{mi}\alpha(x) = u^{jn} = \alpha(z^n)$$

that is, $\alpha(x) = u^{jn - im}$, and finally

$$\alpha(y) = u^{i - (jn - im)}.$$

Therefore all solutions are cyclic. We are thus left with the case $m = 1$.

PROPOSITION 9.2.3. *Let* $\alpha: \Xi^* \to A^*$ *be a noncyclic solution of the equation* (xyx, z^n) *where* $n > 1$. *Then there exist two words* $u, v \in A^*$ *and an integer* $i \geq 0$ *such that*

$$\alpha(x) = (uv)^i u, \qquad \alpha(y) = vu((uv)^{i+1}u)^{n-2} uv, \quad \text{and} \quad \alpha(z) = (uv)^{i+1}u.$$

Proof. Using the same argument as in the discussion of the equation $((xy)^m x, z^n)$ with $m > 1$, we may assume $|\alpha(x)| < |\alpha(z)|$. Then comparing the left factors and then the right factors of length $|\alpha(z)|$ in the equality $\alpha(x)\alpha(y)\alpha(x) = \alpha(z)^n$, we get $\alpha(z) = \alpha(x)w = t\alpha(x)$ for some $w, t \in A^*$. Then by Proposition 1.3.4 of Chapter 1, we have, for some $u, v \in A^*$ and some $i \geq 0$: $\alpha(x) = (uv)^i u$, $\alpha(z) = (uv)^{i+1}u, w = vu$ and $t = uv$, which completes the proof. ∎

We now turn to the main result of this section.

THEOREM 9.2.4. *For all integers* $n, m, p \geq 2$, *the equation* $(x^n y^m, z^p)$ *admits only cyclic solutions.*

Proof. Assume by contradiction that there exist a finite set A, three words $u, v, w \in A^*$ that are not powers of a common word, and integers $n, m, p \geq 2$

such that the equality $u^n v^m = w^p$ holds. Further, we may assume that w is of minimal length. Observe that under these conditions, no two of the three words u, v, w are powers of a common word.

By Proposition 1.3.5 of Chapter 1 the two following inequalities hold:

$$(n-1)|u| < |w| \tag{9.2.1}$$

$$(m-1)|v| < |w| \tag{9.2.2}$$

We can rule out two trivial cases. Assume $p \geq 4$. Then we have

$$|u^n v^m| < 2((n-1)|u| + (m-1)|v|) < 4|w| \leq |w^p|$$

Now if $p = 3$ and $n, m \geq 3$, we obtain:

$$|u^n v^m| < \tfrac{3}{2}((n-1)|u| + (m-1)|v|) < 3|w| \leq |w^p|$$

We are thus left with the following cases:

Case 1. $n = 2$, $m \geq 3$, and $p = 3$ (by symmetry this covers the case $n \geq 3$, $m = 2$, and $p = 3$).
Inequalities (9.2.1) and (9.2.2) yield

$$(m-1)|v| < |w| \quad \text{and} \quad |u| < |w|, \quad \text{i.e.} \quad |w| < |u^2| < 2|w|.$$

Thus, there exist two words $w_1, w_2 \in A^*$ with

$$w_1 w_2 w_1 = u^2, \tag{9.2.3}$$

$$w_2 w_1 w_2 = v^m \tag{9.2.4}$$

and $w_1 w_2 = w$. By Proposition 9.2.3, the first equality implies $w_1 = (ab)^r a$, $w_2 = ba^2 b$ and $u = (ab)^{r+1} a$ for some $a, b \in A^*$ and $r \geq 0$.

Assume first $r \leq 1$; that is, $|w_2| > |w_1|$. Then because of Proposition 9.2.3 applied to Eq. (9.2.4), w_1, w_2 and v are powers of a common word and so are u, v, and w.

Assume next $r \geq 2$, and substitute w_1 and w_2 in Eq. (9.2.4). We get

$$ba(ab)^{r+2} aab = v^m. \tag{9.2.5}$$

Then for some conjugate v' of v we have $(ab)^{r+2} a\, abba = v'^m$. Because of $m \geq 3$, we obtain $|(ab)^{r+2} a| \geq |ab| + |v'|$, which by Proposition 1.3.5 shows that ab and v' are powers of a common word. This yields $ab = (cd)^i$ and $v = (dc)^j$ for some $c, d \in A^*$ and some $i, j \geq 0$. Comparing the two right factors of length $|ab|$ in (9.2.5) we have $ab = (dc)^i$. Thus a, b and therefore u, v, w are powers of a common word.

Case 2. $n = m = 2$ and $p = 3$. As in the former case, for some $w_1, w_2 \in A^*$ we have

$$w_1 w_2 w_1 = u^2 \tag{9.2.6}$$

$$w_2 w_1 w_2 = v^2 \tag{9.2.7}$$

and $w_1 w_2 = w$. By Proposition 9.2.3, the first equality implies $w_1 = (ab)^r a$, $w_2 = ba^2 b$, and $u = (ab)^{r+1} a$, where $a, b \in A^*$ and $r \geqslant 0$. Substituting in Eq. (9.2.7) gives $v^2 = ba(ab)^{r+2} aab$.

Assume first that r is even: $r = 2r'$. Then we obtain $v = ba(ab)^{r'+1} a_1 = a_2(ba)^{r'+1} ab$, where $a = a_1 a_2$ and $|a_1| = |a_2|$. Comparing the two left factors of length $|ba|$ of this last equality, yields $ba_1 a_2 = a_2 ba_1$ —that is, $a_1 = a_2$. Therefore, a and b are powers of a common word, and so are u and v.

Now if r is odd, $r = 2r' + 1$; then we get $v = ba(ab)^r ab_1 = b_2(ab)^r aab$ where $b = b_1 b_2$ and $|b_1| = |b_2|$. Comparing the two left factors of length $|ba|$ in this last equality yields $b_1 b_2 a = b_2 ab_1$ —that is, $b_1 = b_2$. Thus a and b are powers of a common word, and so are u and v.

Case 3. $n, m \geqslant 2$ and $p = 2$. By symmetry, we may assume that there exist $v_1, v_2 \in A^*$ with $w = u^n v_1 = v_2(v_1 v_2)^{m-1}$ where $v = v_1 v_2$. Multiply both members of this last equality on the right by v_1. Then we get $u^n v_1^2 = (v_2 v_1)^m$. Since $|v_2 v_1| = |v| < |w|$, the minimality of $|w|$ implies that u, v_1 and $v_2 v_1$ are powers of a common word. Thus so are u and v.

As a consequence, in all cases, we are led to a contradiction. This completes the proof. ∎

9.3. Equations in Three Unknowns

The solutions of each equation considered in the preceding section are given by a general formula where two types of "parameters" occur: "word parameters" (u and v) and "numerical parameters" (i, j, k, and l). This is true in general for equations in three unknowns. This fact can be formalized using the notion of parametric words (see Lyndon 1960), which we shall now recall.

Denote by $\mathbb{N}[T]$ the semiring of polynomials with coefficients in $\mathbb{N}$ in the commutative indeterminates $T = \{t_1, \ldots, t_k\}$. A *parametric word* over the alphabet Θ is defined as follows:

(i) Every letter in Θ is a parametric word.
(ii) If w_1 and w_2 are parametric words, then $(w_1 w_2)$ is a parametric word.
(iii) If w is a parametric word and P a polynomial in $\mathbb{N}[T]$, then $(w)^P$ is a parametric word.
(iv) All parametric words are obtained inductively applying (i), (ii), (iii).

Given a parametric word w, every assignment of values in a free monoid A^* to the letters of Θ, and of values in $\mathbb{N}$ to the indeterminates $\{t_1, \ldots, t_k\}$ defines a unique word w' in A^*. We will say that w' is the *value* of w (under this assignment). Then we have (see Hmelevskii 1976):

THEOREM 9.3.1. *Given an equation with three unknowns x, y, z there exists a finite family $\{(u_i, v_i, w_i)\}_{1 \leqslant i \leqslant r}$ of triples of parametric words over an alphabet with two letters, such that for each noncyclic solution α: $\Xi^* \to A^*$ we have*

$$\alpha(x) = u_i', \qquad \alpha(y) = v_i', \qquad \alpha(z) = w_i',$$

where for some $1 \leqslant i \leqslant r$, u_i', v_i', and w_i' are the values of u_i, v_i and w_i under the same assignment.

We shall express this fact by saying that the equations in three unknowns are *parametrizable*. Actually Hmelevskii shows more than that. He proves that we can associate with each triple a finite family of linear inequalities in the indeterminates $t_1, \ldots, t_k$, in such a way that the family of triples, together with these inequalities, characterizes all noncyclic solutions of the given equation. Moreover his proof gives an effective construction of the triples and inequalities in question. However, in the same work (1976), he points out that the equation (xyz, ztx) in four unknowns is not parametrizable.

9.4. Rank of an Equation

In the preceding section the fact was mentioned that equations in more than three unknowns are not parametrizable—that is, that there does not exist a finite collection of formulas representing all their solutions. As an illustration of what happens, let us consider the equation (xyz, ztx), for which it was recalled at the end of Section 9.3 that such a finite collection does not exist.
Consider the particular solution:

$$\alpha(x) = ab, \qquad \alpha(y) = babca, \qquad \alpha(z) = ab^2ab, \qquad \alpha(t) = ca^2b^2,$$

where a, b, c are letters of an alphabet A. The basis of the free hull containing $\alpha(x)$, $\alpha(y)$, $\alpha(z)$, and $\alpha(t)$, consists of the three words $u = ab$, $v = b$, and $w = ca$. Now if we no longer consider u, v, w as words over the alphabet A but, rather, as letters of a new alphabet B, then the morphism:

$$\beta(x) = u, \qquad \beta(y) = vuw, \qquad \beta(z) = uvu, \qquad \beta(t) = wuv$$

obtained after substituting u, v, and w in $\alpha(x)$, $\alpha(y)$, $\alpha(z)$, and $\alpha(t)$, again satisfies the equation (xyz, ztx). Thus u, v, w may be viewed as word parameters of a parametric word in the same way as in the preceding section. Consequently, the rank of a solution appears as the number of word parameters occurring in the most general formula of which it is a value under some assignment.

Therefore, since given an equation there is no guarantee that its solutions can be represented by a finite collection of parametric words as defined in Section 9.3, we will ask for the maximum number of word parameters occurring in the parametric words obtained as indicated in the foregoing example.

We are thus led to define the *rank* of an equation (e, e'), written: rank(e, e'), as the maximum rank of its solutions. In particular we have rank$(e, e') \leqslant 1$ iff (e, e') has only cyclic solutions. Henceforth the study of equations here will be focused on determining their rank. We first show that instead of considering all solutions we can restrict ourselves to special solutions called *principal solutions*.

A solution $\alpha: \Xi^* \to A^*$ of the equation (e, e') is *principal* if for all solutions $\beta: \Xi^* \to B^*$ dividing it, that is, for which there exists a nonerasing morphism $\theta: B^* \to A^*$ with $\alpha = \theta \circ \beta$, we have $\alpha \approx \beta$. Remember that because of the convention of Section 9.1, we do not distinguish between two equivalent principal solutions.

Clearly, any solution can be divided by some principal solution. Indeed, if $\beta: \Xi^* \to B^*$ is principal, there is nothing to prove. Otherwise, β can be divided by some nonequivalent solution $\alpha: \Xi^* \to A^*$, which means that there exists a nonerasing morphism $\theta: A^* \to B^*$ verifying $\beta = \theta\alpha$ and $|\theta(x)| > 1$ for some $x \in \Xi$. Then we obtain

$$\sum_{x \in \Xi} |\alpha(x)| < \sum_{x \in \Xi} |\beta(x)|.$$

Thus after finitely many steps, we get a principal solution dividing β. Actually we shall see in the next section that there exists a unique principal solution dividing a given solution.

The importance of principal solutions is due to the fact that according to Proposition 9.1.2, the rank of (e, e') is equal to the maximum rank of all its principal solutions. Notice further that the rank of a principal solution $\alpha: \Xi^* \to A^*$ equals the cardinality of A: rank $(\alpha) = $ Card A.

Example 9.4.1. The principal solutions $\alpha: \Xi^* \to A^*$ of the equation (xy, yz) are of one of the following types:

1. $\alpha(x) = \alpha(y) = \alpha(z) = 1$.
2. $\alpha(x) = \alpha(z) = 1, \alpha(y) = a$ with $A = \{a\}$.

3. $\alpha(x) = \alpha(z) = a,$ $\alpha(y) = a^p$ with $A = \{a\}$, and $p \geqslant 0$.

4. $\alpha(x) = ab,$ $\alpha(y) = (ab)^p a,$ $\alpha(z) = ba$ with
      $A = \{a, b\}$ and $p \geqslant 0$.

Thus rank$(xy, yz) = 2$.

Finally, the study of the rank of an equation reduces to the study of the rank of the nonerasing solutions of certain equations as follows:

Let $\alpha: \Xi^* \to A^*$ be a nontrivial solution of the equation (e, e'). Then $\Xi' = \{x \in \Xi \mid \alpha(x) \neq 1\}$ is nonempty. Denote by Π the projection of Ξ^* into Ξ'^* defined by

$$\Pi(x) = \begin{cases} x & \text{if} \quad x \in \Xi' \\ 1 & \text{if} \quad x \in \Xi \setminus \Xi' \end{cases}$$

Then the morphism $\alpha': \Xi'^* \to A^*$ defined by $\alpha'(x) = \alpha(x)$ for all $x \in \Xi'$ is a nonerasing solution of the equation $(\Pi(e), \Pi(e'))$ and we have the following:

PROPOSITION 9.4.2. *The solution α of the equation (e, e') is principal iff the solution α' of the equation $(\Pi(e), (\Pi(e')))$ is itself principal.*

The proof is left to the reader.

9.5 Fundamental Solutions

The purpose of this section is to prove the uniqueness of the principal solution dividing a given solution. This result can be established with the help of different methods available in the literature (such as that in Lentin 1972a). We shall adapt here, in the case of free monoids, a method used in the theory of equations in free groups and based on the Nielsen transformations. This leads to the notion of fundamental solutions, whose equivalence to the principal solutions we prove. The reader is referred to Lyndon (1959) for the corresponding result for free groups.

It is convenient to extend the notation alph to any subset. For each $X \subseteq \Xi^*$ we define:

$$\text{alph } X = \bigcup_{x \in X} \text{alph } x.$$

Let now x and x' be two arbitrary distinct letters of Ξ. The following terminology is borrowed from Lyndon (1959):

We denote by $\varphi_{xx'}$ the morphism of Ξ^* into itself, which takes x' to xx' and which leaves all other letters invariant:

$$\varphi_{xx'}(y) = \begin{cases} y & \text{if } y \in \Xi \setminus \{x'\} \\ xx' & \text{if } y = x' \end{cases}$$

Then $\varphi_{xx'}$ is injective. More precisely its extension to the free group generated by Ξ defines an automorphism of this group, which is a Nielsen transformation.

We denote by $\varepsilon_{xx'}$ the morphism of Ξ^* into itself, which takes x' to x and which leaves all other letters invariant:

$$\varepsilon_{xx'}(y) = \begin{cases} y & \text{if } y \in \Xi \setminus \{x'\} \\ x & \text{if } y = x' \end{cases}$$

We shall now associate with certain pairs of words $(e, e') \in \Xi^* \times \Xi^*$, some of the transformations just defined. *We do not assume any longer that alph$(ee') = \Xi$.*

Suppose neither e nor e' is a left factor of the other: $e = gxh$ and $e' = gx'h'$, where $g, h, h' \in \Xi^*$, $x, x' \in \Xi$ and $x \neq x'$. Then we say that the two morphisms $\varphi_{xx'}$ and $\varphi_{x'x}$ are the two *regular elementary transformations attached* to (e, e') and that the morphism $\varepsilon_{xx'}$ is the *singular elementary transformation attached to (e, e').*

More generally, a *transformation attached* to (e, e') is any product $\varphi_n \ldots \varphi_1$ where each $1 \leq i \leq n$, φ_i is an elementary transformation attached to $(\varphi_{i-1} \ldots \varphi_1(e), \varphi_{i-1} \ldots \varphi_1(e'))$. When $e = e'$, the unique transformation attached to (e, e') is, by definition, the identity over Ξ^*.

Consider a transformation φ attached to (e, e') and satisfying $\varphi(e) = \varphi(e')$. We set $\Gamma = \text{alph}(ee')$ and $\Delta = \text{alph}\varphi(\Gamma)$. If we identify φ with the total morphism it defines from Γ^* into Δ^*, then φ can be viewed as a solution of the equation (e, e'). Such a solution is called *fundamental*.

Example 9.5.1. Consider the equation (xyz, xzx)

Pairs of words	Attached elementary transformations
$(e_1, e_1') = (xyz, xzx)$	$\varphi_1 = \varphi_{zy}$
$(e_2, e_2') = (xzyz, xzx)$	$\varphi_2 = \varphi_{yx}$
$(e_3, e_3') = (yxzyz, yxzyx)$	$\varphi_3 = \varepsilon_{zx}$
$(e_4, e_4') = (yzzyz, yzzyz)$	

Thus $\varphi_3 \varphi_2 \varphi_1 = \varphi$ is a fundamental solution of (xyz, xzx). It is defined by

$$\varphi(x) = yz \qquad \varphi(y) = zy \qquad \varphi(z) = z.$$

The next proposition shows that with the identification just defined, a solution is principal iff it is fundamental:

PROPOSITION 9.5.2. *Each nonerasing solution* α: $\Xi^* \to A^*$ *of the equation* (e, e') *has a unique factorization* $\alpha = \theta\varphi_n \ldots \varphi_1$ *where* $\varphi_n \ldots \varphi_1$ *is a factorization of a fundamental solution of* (e, e') *into elementary transformations and where for all* $x \in \Xi$ *we have* $\theta(x) \neq 1$ *iff* $x \in alph\varphi_n \ldots \varphi_1(\Xi)$.

Proof. Let g be the largest common left factor of e and e': $e = gf$ and $e' = gf'$. We shall proceed by induction on $m = \text{Card}(\text{alph}(ff'))$.

Assume $m = 0$. Then we have $e = g = e'$. The unique factorization of α is obtained by setting $\theta = \alpha$, $n = 1$, and by taking the identity morphism for φ_1.

If $m > 0$, then necessarily we have $e = gxh$ and $e' = gx'h'$, where $f = xh$, $f' = x'h'$, $x, x' \in \Xi$, and $x \neq x'$. Thus we have

$$\alpha(e) = \alpha(g)\alpha(x)\alpha(h) = \alpha(g)\alpha(x')\alpha(h') = \alpha(e') \qquad (9.5.1)$$

For a fixed m, we proceed by induction on:

$$\sigma_\alpha = \text{Min}\{S||\alpha(x)| - |\alpha(x')||S,\} \quad \text{with} \quad S = \sum_{x \in \Xi} |\alpha(x)|.$$

Let us first consider the case $\sigma_\alpha = 0$; that is, by Eq. (9.5.1):

$$\alpha(x) = \alpha(x'). \qquad (9.5.2)$$

If there exists a factorization satisfying the proposition, then necessarily the elementary transformation φ_1 is singular. Indeed, assume by contradiction that we have for instance: $\varphi_1 = \varphi_{x'x}$. Then we get

$$\alpha(x) = (\theta\varphi_n \ldots \varphi_2(x'))(\theta\varphi_n \ldots \varphi_2(x)) = (\theta\varphi_n \ldots \varphi_1(x'))(\theta\varphi_n \ldots \varphi_2(x))$$
$$= \alpha(x')(\theta\varphi_n \ldots \varphi_2(x)). \qquad (9.5.3)$$

Because of the hypothesis, this shows that $\alpha(x')$ is a proper left factor of $\alpha(x)$, which contradicts (9.5.2). Setting $\varphi_1 = \varepsilon_{xx'}$, we get $\alpha = \alpha\varphi_1$, which yields $\alpha\varphi_1(e) = \alpha\varphi_1(e')$. Consider $\varphi_1(e) = g_1h_1$ and $\varphi_1(e') = g_1h_1'$, where g_1 is the longest common left factor of $\varphi_1(e)$ and $\varphi_1(e')$. Then we obtain: $\text{Card}(\text{alph}(h_1h_1')) \leq m - 1$, which by the induction hypothesis implies the existence of a unique factorization, $\alpha = \theta\varphi_n \ldots \varphi_2$, satisfying the conditions of the proposition for the solution α of the equation $(\varphi_1(e), \varphi_1(e'))$. Then $\alpha = \theta\varphi_n \ldots \varphi_1$ is the required factorization.

Consider now the case $\sigma_\alpha > 0$; that is, $\sigma_\alpha = S$. Without loss of generality we may assume that $|\alpha(x)| > |\alpha(x')|$. Because of equality (9.5.1), we get $\alpha(x) = \alpha(x')u$ for some $u \in A^+$. As in the preceding case, it can be verified that if α admits a factorization as in the proposition, then necessarily

$\varphi_1 = \varphi_{xx'}$. Let $\beta\colon \Xi^* \to A^*$ be the morphism defined by $\beta(x) = u$ and $\beta(y) = \alpha(y)$ if $y \in \Xi \setminus \{x\}$. Then we have $\alpha = \beta\varphi_1$ and thus $\beta\varphi_1(e) = \beta\varphi_1(e')$. Furthermore we obtain

$$\sigma_\beta \leqslant \sum_{x \in \Xi} |\beta(x)| = S - |\alpha(x')| < S.$$

By the induction hypothesis there exists a unique factorization $\beta = \theta\varphi_n \ldots \varphi_2$ satisfying the proposition. It then suffices to consider the factorization $\alpha = \theta\varphi_n \ldots \varphi_1$. ∎

The preceding result shows that each nonerasing solution can be divided by a unique fundamental solution that is also a principal solution. Conversely it shows that each principal nonerasing solution equals some fundamental solution. The existence and uniqueness of the principal solution dividing a given solution can, with the help of Proposition 9.4.2, be extended to all (not necessarily nonerasing) solutions. Thus we have the following:

THEOREM 9.5.3. *Each solution of a given equation can be divided by some unique principal solution.*

Let now $\alpha\colon \Xi^* \to A^*$ be a principal nonerasing solution of the equation (e, e'). As observed in Section 9.4, we have rank $\alpha = $ Card A. Moreover, if $\varphi_n \ldots \varphi_1$ is the factorization of α into elementary transformations, we have Card $A = \mathrm{Card}(\mathrm{alph}(\varphi_n \ldots \varphi_1(ee')))$. Thus we can state:

COROLLARY 9.5.4. *Let $\varphi_n \ldots \varphi_1$ be the factorization into elementary transformations of a principal nonerasing solution α of (e, e'). Then we have*

$$\mathrm{rank}\ \alpha = \mathrm{Card}(\mathrm{alph}(ee')) - k$$

where k is the number of singular transformations occurring in $\varphi_n \ldots \varphi_1$.

Since all regular transformations are injective, in the factorization $\varphi_n \ldots \varphi_1$ of a principal nonerasing solution there always occurs at least one singular transformation. Therefore, the rank of any nonerasing principal solution of the equation (e, e'), with $e \neq e'$, is inferior to the number $\mathrm{Card}(\mathrm{alph}(ee'))$. This can be extended without any difficulty to all (not necessarily nonerasing) principal solutions. We thus get the "equational" version of the defect theorem (1.2.5) of Chapter 1:

THEOREM 9.5.5. *Let $(e, e') \in \Xi^* \times \Xi^*$ be an equation such that $e \neq e'$. Then $\mathrm{rank}(e, e') < \mathrm{Card}(\mathrm{alph}(ee'))$.*

9.6.   The Graph Associated with an Equation

Proposition 9.5.2 shows how we get a principal solution of a given equation by applying successive transformations to the equation. This suggests the idea of defining an oriented graph whose vertices are the different equations obtained and whose edges are labeled by the elementary transformations applied at each step (see Lentin (1972a)).

Formally, denote by E the subset of $\Xi^* \times \Xi^*$ consisting of the pair $(1, 1)$ and of all pairs (f, f') of words such that f and f' are nonempty and start with a different letter. We consider E as the set of vertices of an oriented graph G, an edge joining the vertex (f, f') to the vertex (g, g') iff there exists an elementary transformation $\varphi : \Xi^* \to \Xi^*$ attached to (f, f') and satisfying $\varphi(f) = hg$ and $\varphi(f') = hg'$, where h is the longest common left factor of $\varphi(f)$ and $\varphi(f')$. Such an edge is labeled by φ.

Consider now an equation $(e, e') \in \Xi^* \times \Xi^*$ and assume e and e' are two nonempty words starting with a different letter (this is no restriction of generality). Then by Proposition 9.5.2, each principal nonerasing solution α is represented in G by a unique path joining (e, e') to $(1, 1)$. The labels in the path give the factorization of α into elementary transformations. Conversely, each such path gives rise to a principal solution whose rank can be calculated using Corollary 9.5.4.

The *graph associated with* (e, e') is defined as the subgraph G' of G whose vertices are all pairs $(f, f') \in \Xi^* \times \Xi^*$ such that there exists, in G, a path going from (e, e') to $(1, 1)$ via (f, f').

Example 9.6.1. The graph associated with the equation (xy, yz) is shown in Figure 9.1. The graph associated with the equation (xyz, zyx) is shown in Figure 9.2.

By Corollary 9.5.4, the rank of the equation (e, e') is equal to $\mathrm{Card}(\mathrm{alph}(ee')) - k$, where k is the minimum, over all paths going from

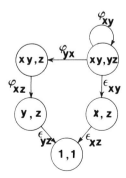

Figure 9.1. Graph associated with the equation (xy, yz).

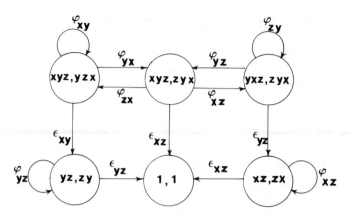

Figure 9.2. Graph associated with the equation (xyz, zyx).

(e, e') to $(1, 1)$, of the number of edges labeled by a singular transformation in such a path.

Example 9.6.1. (continued). The rank of the equation (xy, yz) equals $3-1=2$. A solution of rank 2 is given, for example by the path φ_{yx}, φ_{xz}, ε_{yz} defining the morphism:

$$\varphi(x)=yx \qquad \varphi(y)=y \qquad \varphi(z)=xy$$

Example 9.6.2. (continued). The rank of the equation (xyz, zyx) equals $3-1=2$. For instance $\varphi(x)=\varphi(z)=x$ and $\varphi(y)=y$ defines a solution of rank 2 corresponding to the path labeled by ε_{xz}.

When the graph associated with an equation is finite, its rank can be computed effectively. Using, for any $f\in\Xi^*$, the following notation:

$$\|f\|=\max\{|f|_x \mid x\in\Xi\}$$

we can state a sufficient condition for the graph to be finite (see Lentin 1972b):

PROPOSITION 9.6.3. *Assume the two following conditions are satisfied:*

(i) *for each* $x\in\Xi$: $|e|_x|e'|_x\leqslant1$;
(ii) $\max\{\|e\|, \|e'\|\}\leqslant2$.

Then the graph associated with (e, e') *is finite.*

9.7.   Quadratic Equations

An equation (e, e') is *quadratic* if for all $x \in \Xi$ we have $|e|_x = |e'|_x \leqslant 1$. We write $e = x_1 \cdots x_p, e' = x_1' \cdots x_p'$, and we say that (e, e') is *irreducible* if for no integer $1 \leqslant i < p, (x_1 \cdots x_i, x_1' \cdots x_i')$ is a quadratic equation—that is, if $\{x_1, \ldots, x_i\} = \{x_1', \ldots, x_i'\}$ holds only if $i = p$.

Using Corollary 9.5.4 we can compute the rank of such an equation via the graph introduced in the preceding section. In fact, for such a special type of equation, there exists a formula giving directly the rank.

Indeed, consider the quadratic equation

$$(x_1 \cdots x_p, x_1' \cdots x_p'),$$

and assume

$$\Xi = \{x_i\}_{1 \leqslant i \leqslant p} = \{x_i'\}_{1 \leqslant i \leqslant p}.$$

Denote by τ (resp. τ') the bijection of $\Xi \setminus \{x_1\}$ onto $\Xi \setminus \{x_p'\}$ (resp. $\Xi \setminus \{x_p'\}$ onto $\Xi \setminus \{x_1\}$) defined for all $1 < i \leqslant p$ (resp. $1 \leqslant i < p$) by

$$\tau(x_i) = \begin{cases} x_p & \text{if } x_{i-1} = x_p' \\ x_{i-1} & \text{otherwise} \end{cases} \qquad \tau'(x_i') = \begin{cases} x_1' & \text{if } x_{i+1}' = x_1 \\ x_{i+1}' & \text{otherwise} \end{cases}$$

If we denote by σ the permutation of $\Xi \setminus \{x_p'\}$ equal to $\tau \cdot \tau'$, then we have (Lentin 1972b):

THEOREM 9.7.1. *Assume* $(x_1 \cdots x_p, x_1' \cdots x_p')$ *is an irreducible quadratic equation. Then its rank is equal to:*

$$\frac{-1 + p + z(\sigma)}{2}$$

where $z(\sigma)$ is the number of cycles of the permutation σ.

This result, based on several lemmas involving permutation groups, cannot be treated here. Notice that a similar result has since been established for quadratic equations in free groups by Piollet (1975).

Example 9.7.2. Consider the quadratic equation $(x_1 x_2 x_3 x_4, x_2 x_3 x_4 x_1)$. We compute:

$$\tau(x_2) = x_4, \qquad \tau(x_3) = x_2, \qquad \tau(x_4) = x_3;$$
$$\tau'(x_2) = x_3, \qquad \tau'(x_3) = x_4, \qquad \tau'(x_4) = x_2;$$
$$\sigma = (x_2)(x_3)(x_4).$$

Thus the rank equals $(-1+4+3)/2 = 3$. This is the rank, for instance, of the principal solution $\varphi: \Xi^* \to A^*$ defined by

$$\varphi(x_1) = abc, \qquad \varphi(x_2) = abca, \qquad \varphi(x_3) = bcab, \qquad \varphi(x_4) = cabc,$$

where $A = \{a, b, c\}$.

9.8. Related Theories

The notion of an equation with constants is obtained by adding to the alphabet Ξ of unknowns, a disjoint alphabet Γ of constants. Setting $\Delta = \Xi \cup \Gamma$, an *equation with constants* is any pair (f, f') of words in Δ^*. A *solution* of such an equation is a morphism α of the free monoid Δ^* into the free monoid of constants Γ^* such that $\alpha(a) = a$ for all $a \in \Gamma$, *satisfying* (f, f')— that is, verifying $\alpha(f) = \alpha(f')$. (For the corresponding concept for free groups, see Lyndon and Schupp 1977.)

Example 9.8.1. Consider $\Xi = \{x, y\}$, $\Gamma = \{a, b\}$ and the equation $(xbacy, ycabx)$. Then for each integer $i \geq 0$, the morphism $\alpha: \Delta^* \to \Gamma^*$, defined by

$$\alpha(x) = (ab)^i a \qquad \alpha(y) = (ab)^{i+1} a \qquad \alpha(a) = a \qquad \alpha(b) = b$$

is a solution of the equation.

Example 9.8.2. Consider $\Xi = \{x\}$, $\Gamma = \{a, b\}$, and the equation (ax, xb). Then by Proposition 1.3.4 of Chapter 1, this equation cannot be satisfied.

Makanin (1977) proved that given an equation with constants, it can be decided whether it has a solution. Very intuitively, he shows that given such an equation, an integer N can be computed effectively, with the property that if the equation can be satisfied, then after at most N particular transformations of the original equation, one obtains an equation where the alphabet of constants is reduced to one letter. This result has a direct consequence for equations without constants. Indeed it implies that the rank of such equations can effectively be computed (cf. Pecuchet 1981).

Another theory dealing with equations is the theory of test sets.

Given two morphisms α, β of Ξ^* into a free monoid A^*, their *equality set* is the set of words in Ξ^* on which they agree—that is, the set defined by the following (cf. Rozenberg and Salomaa 1980):

$$E(\alpha, \beta) = \{w \in \Xi^* \mid \alpha(w) = \beta(w)\}$$

Consider now any subset L of Ξ^*. A *test set* for L is a *finite* subset $L_0 \subseteq L$ such that for any pair of morphisms $\alpha, \beta: \Xi^* \to A^*$, the two morphisms have the same restriction to L, once they have the same restriction to L_0: $L_0 \subseteq E(\alpha, \beta)$ implies $L \subseteq E(\alpha, \beta)$.

It has been conjectured by Ehrenfeucht that every subset possesses a test set. This has been positively proved in the case when Ξ consists of two letters (cf. Culik and Salomaa 1980, Ehrenfeucht et al. 1982). However, in the general case, the problem is still open.

The problem can be reformulated in terms of systems of equations. Let $\overline{\Xi}$ be a disjoint copy of Ξ, and for each $x \in \Xi$ denote by $\bar{x}$ the corresponding letter in $\overline{\Xi}$. This application extends to a unique isomorphism $w \to \overline{w}$ of Ξ^* onto $\overline{\Xi}^*$. With every pair $\alpha, \beta \colon \Xi^* \to A^*$ of morphisms, we associate the morphism $\alpha * \beta \colon (\Xi \cup \overline{\Xi})^* \to A^*$ defined for all $x \in \Xi$ by

$$\alpha * \beta(x) = f(x) \quad \text{and} \quad \alpha * \beta(\bar{x}) = g(x)$$

Given a subset $L \subseteq \Xi^*$, we denote by $\mathfrak{L}$ the system of all equations $(w, \overline{w})$ where $w \in L$. Since any two morphisms $\alpha, \beta \colon \Xi^* \to A^*$ have the same restriction to L iff $\alpha * \beta$ is a solution of the system $\mathfrak{L}$, the problem of finding a test set L_0 for L amounts to finding a *finite* subsystem $\mathfrak{L}_0 \subseteq \mathfrak{L}$ having the same solutions as $\mathfrak{L}$.

More generally, we could ask under which conditions a system of equations is equivalent to a finite subsystem.

In this chapter the problem of finding solutions of equations has been discussed. Let us conclude the chapter with a few words on the dual problem, to wit: Given a morphism $\alpha \colon \Xi^* \to A^*$, what is the family of equations it satisfies?

The problem was considered by Spehner (1976, 1981) in the case when Ξ consists of three letters. He shows that the family of equations a given morphism satisfies is one of finitely many explicitly given types.

Finally, the dual problem has strong connections with some famous problems such as the "Post correspondence problem."

Recall that given a finite family of pairs $\{(u_i, v_i)\}_{1 \leqslant i \leqslant n}$ of words in the free monoid A^*, the Post correspondence problem consists in asking whether there exists a sequence of integers $i_1, \ldots, i_r$, where $r > 0$ and $1 \leqslant i_k \leqslant n$ for all $1 \leqslant k \leqslant r$, such that the following holds:

$$u_{i_1} \ldots u_{i_r} = v_{i_1} \ldots v_{i_r}$$

Example 9.8.3. Let $A = \{a, b\}$ and consider the two pairs of words (u_1, v_1) and (u_2, v_2) where $u_1 = a$, $v_1 = ababa$, $u_2 = bababab$, and $v_2 = b$. Then we obtain $u_1 u_2 u_1 u_2 u_1 = v_1 v_2 v_1 v_2 v_1$, which shows that in this case the Post correspondence problem can be answered positively.

We set $\Xi = \{x_1, \ldots, x_n\}$ and $\overline{\Xi} = \{\bar{x}_1, \ldots, \bar{x}_n\}$ and we assume that Ξ and $\overline{\Xi}$ are disjoint. Let $w \to \overline{w}$ be the natural isomorphism between Ξ^* and $\overline{\Xi}^*$, and let α be the morphism of $(\Xi \cup \bar{\Xi})^*$ into A^* defined for each $1 \leqslant i \leqslant n$ by

$$\alpha(x_i) = u_i \quad \text{and} \quad \alpha(\bar{x}_i) = v_i$$

Then the Post correspondence problem amounts to asking whether α satisfies an equation of the form $(w, \bar{w})$.

Problems

Section 9.1

9.1.1. A morphism α: $\Xi^* \to A^*$ is *elementary* if for every factorization $\alpha = \theta\beta$ into two morphisms β: $\Xi^* \to B^*$ and θ: $B^* \to A^*$ we have Card $B \geqslant$ Card A.

Using the defect theorem (1.2.5) of Chapter 1, show that each elementary morphism is injective. Give an example of an injective morphism that is not elementary.

Show that when Ξ consists of two letters every injective morphism is elementary (see Rozenberg and Salomaa 1980).

*9.1.2. Let α: $\Xi^* \to A^*$ be an elementary morphism (see the preceding problem). Set

$$K = \sum_{x \in \Xi} (|\alpha(x)| - 1).$$

Show that if the words $x, x' \in \Xi, w, w' \in \Xi^*$ satisfy $|\alpha(xw)| \geqslant K$ and $\alpha(xw)u = \alpha(x'w')$ for some $u \in A^*$, then $x = x'$. Exhibit an injective morphism that does not satisfy this property. Exhibit a morphism that is not elementary and that satisfies this property. Show that this result is a generalization of Proposition 1.3.5 of Chapter 1 (see Rozenberg and Salomaa 1980).

Section 9.2

9.2.1. Solve the equation (xy^n, z^m).
9.2.2. Solve the equation $((xy)^n x, (zt)^n z)$.

Section 9.3

9.3.1. Find the parametric words, some values of which are solutions of the equation (xyx, zxz). Do the same for the equation $(x^2 y, y^2 z)$.

Section 9.4

9.4.1. Characterize all the principal solutions of the equation (e, e') where $e = e'$.

9.4.2. Give an example of an equation in n unknowns, of rank $n - 1$, all the principal nonerasing solutions of which have a rank equal to 1.

In the next two problems, we consider a fixed equation (e, e') in the unknowns $\Xi = \{x_1, \ldots, x_p\}$ and for $1 \leq i \leq p$ we set

$$\Delta_i = |e|_{x_i} - |e'|_{x_i}.$$

9.4.3. Give a necessary and sufficient condition on the Δ_i's for (e, e') to have a nonerasing solution.

9.4.4. Let $\alpha \colon \Xi^* \to A^*$ be a cyclic morphism, and for every $1 \leq i \leq p$ set $\alpha(x_i) = u^{r_i}$, where u is a fixed element of A^* and $r_i \geq 0$. Then α can be identified with the p-uple $(r_1, \ldots, r_p) \in \mathbb{N}^p$. Further, α is a solution of (e, e') iff the linear equation

$$\sum_{1 \leq i \leq p} \Delta_i \rho_i = 0 \tag{1}$$

in the unknowns ρ_i is satisfied when assigning the value r_i to ρ_i for all $1 \leq i \leq p$.

Denote by $\leq$ the product ordering over $\mathbb{N}^p$ defined by: $z \leq t$ iff there exists w in $\mathbb{N}^p$ with $z + w = t$. Show that every solution of (1) is a linear combination, with coefficients in $\mathbb{N}$, of the minimal solutions. Show that there are finitely many minimal solutions (see Problem 6.1.2) Application: Give all the cyclic solutions of the equation $(x_1 x_2 x_1^2 x_2 x_3, x_2 x_3^4 x_1)$.

Section 9.5

9.5.1. Show that if two principal solutions $\alpha \colon \Xi^* \to A^*$ and $\beta \colon \Xi^* \to B^*$ of the same equation verify $|\alpha(x)| = |\beta(x)|$ for all $x \in \Xi$, then they are equivalent.

9.5.2. Verify directly that all the principal nonerasing solutions $\alpha \colon \Xi^* \to A^*$ of the equation (xy, yx) are defined by

$$\alpha(x) = a^n \qquad \alpha(y) = a^m, \quad \text{where } A = \{a\} \quad \text{and} \quad (n, m) = 1$$

(n, m) denotes the greatest common divisor of n and m.

Using Proposition 9.5.2, exhibit a bijection of the set $\mathbb{Q}_+$ of nonnegative rational numbers onto the free monoid generated by two elements.

9.5.3. Let $\alpha: \Xi^* \to A^*$ be a solution of the equation (e, e') and $A' \subseteq A$ be the subset of all letters $a \in A$ appearing as the first letter of $\alpha(x)$ for some $x \in \Xi$. Show that rank $(e, e') \geqslant \operatorname{Card} A'$.

9.5.4. Using the elementary transformations, show that for every principal solution $\alpha: \Xi^* \to A^*$, each letter $a \in A$ appears as the last letter of $\alpha(x)$ for some $x \in \Xi$. Show by symmetry, that "last" can be replaced by "first" in the preceding statement.

9.5.5. Using the preceding problem show that the equation $(xyxzyz, zxxyzx)$ admits only cyclic solutions.

9.5.6. Extend the notion of principal solutions to the solutions of a finite system $\mathbb{S}$ of equations as follows: A solution $\alpha: \Xi^* \to A^*$ of $\mathbb{S}$ is *principal* iff all solutions $\beta: \Xi^* \to B^*$ dividing α are equivalent to α. The *rank* of $\mathbb{S}$ is defined as the maximum rank of its principal solutions.

Show that every solution of $\mathbb{S}$ can be divided by some unique principal solution. Show that the result of Problem 9.5.4 still holds. Give a general condition on the two equations (e_1, e_1') and (e_2, e_2') that ensures that the rank of the system $\mathbb{S} = \{(e_1, e_1'), (e_2, e_2')\}$ is less than or equal to $\operatorname{Card}(\Xi) - 2$.

Section 9.6

9.6.1. Draw the graphs associated with the equations (xyz, tyx), (xyz, ztx), and $((x^2y, z^2), (x^2y^2, z^2))$.

9.6.2. Consider the equation (e, e') with $e = x_1x_2x_4x_1$ and $e' = x_3x_1^2x_5$. Show that in the graph associated with the equation, there exist arbitrarily long paths going from the vertex (e, e') to the vertex $(1, 1)$ and never going twice through the same vertex.

Section 9.7

9.7.1. Show that the rank of the quadratic equation $(x_1x_2 \cdots x_n, x_n \cdots x_2x_1)$ is equal to $[n + 1]/2$, where $[p]$ denotes the largest integer less than or equal to p.

9.7.2. Show that for each quadratic equation in more than two unknowns, it is possible to exhibit a solution of rank greater than 1.

Section 9.8

9.8.1. Show that the satisfiability of an equation with two constants reduces to the satisfiability of a system of diophantine equations (*Hint*: Let $\Gamma = \{a, b\}$, $\Xi = \{x_1, \ldots, x_p\}$ and let $\mathbb{N}[T]$ to be semiring of polynomials with coefficients in $\mathbb{N}$, in the commutative indeterminates $T = \{z_{ij} | 1 \leqslant i \leqslant p, 1 \leqslant j \leqslant 4\}$. Consider the morphism Ψ

of $(\Gamma \cup \Xi)^*$ into the multiplicative monoid of all 2×2-matrices with entries in $\mathbb{N}[T]$, defined by

$$\Psi(a) = \begin{bmatrix} 1 & 1 \\ 0 & 1 \end{bmatrix}, \qquad \Psi(b) = \begin{bmatrix} 1 & 0 \\ 1 & 1 \end{bmatrix} \quad \text{and} \quad \Psi(x_i) = \begin{bmatrix} z_{i1} & z_{i2} \\ z_{i3} & z_{i4} \end{bmatrix}$$

for every $1 \le i \le p$. Then use the fact that $\Psi(a)$ and $\Psi(b)$ freely generate the monoid of matrices with determinant equal to 1 and with entries in $\mathbb{N}$.

9.8.2. Let $\mathbb{S}$ be a finite system of equations with constants. Let the alphabet of constants contain at least two elements. Show that there exists a single equation (e, e') with the same sets of unknowns and constants that is equivalent to $\mathbb{S}$ — that is, such that every morphism satisfies $\mathbb{S}$ iff it satisfies (e, e').

9.8.3. Show that the equality set of two morphisms is a free submonoid generated by a prefix (see Chapter 1).

9.8.4. Let $\alpha: \Xi^* \to A^*$ be an injective morphism and $\beta: \Xi^* \to A^*$ be a cyclic morphism. Show that there exists a word $w \in \Xi^*$ such that $E(\alpha, \beta) = w^*$.

9.8.5. Using Makanin's result, show that given any word $w \in \Xi^*$, it can be decided whether there exist two different morphisms α, β, one of which is not cyclic, satisfying $\alpha(w) = \beta(w)$ (see Culik and Karhumäki 1980).

9.8.6. Consider the equation $(abx, x\tilde{a}\tilde{b})$ where x is the unique unknown and a and b are two arbitrary words over the alphabet Γ of constants ($\tilde{a}$ and $\tilde{b}$ are the reverse words of a and b (see Chapter 1)). Show that the equation has solutions iff ab is the product of two palindromes—that is, iff $ab = cd$ where $c = \tilde{c}$ and $d = \tilde{d}$ (see De Luca 1979).

*9.8.7. Let $\alpha, \beta: \Xi^* \to A^*$ be two injective morphisms such that $\alpha(\Xi)$ and $\beta(\Xi)$ are prefixes (see Chapter 1). Show that there exists an integer $n > 0$ with the property that for all $u, v \in \Xi^*$, $\alpha(uv) = \beta(uv)$ implies $||\alpha(u)| - |\beta(u)|| < n$ (*Hint*: Denote by M the maximum length of the words in $\alpha(\Xi)$ and $\beta(\Xi)$ and observe that for all $u, v, w \in \Xi^*$, $x, y \in \Xi$ satisfying $\alpha(uxv) = \beta(uxv)$ and $\alpha(uyw) = \beta(uyw)$, the inequality $||\alpha(u)| - |\beta(u)|| \ge M$ implies $x = y$.)

Using Problem 9.1.2, show that the preceding result is still valid when α and β are elementary morphisms (see Rozenberg and Salomaa 1980).

CHAPTER 10

Rearrangements of Words

10.1. Preliminaries

When enumerating permutations of finite sequences according to certain patterns (such as with a given number of descents, with a fixed up–down sequence, or with given positions for the maxima) one is frequently led to transfer the counting problem to another class of permutations for which the problem is straightforward or at least easier. Of course there is no general rule to make up those transfers, but we have at our disposal several natural algorithms. The purpose of this chapter is to describe those algorithms and mention several applications.

The typical set-up for describing those algorithms is the following. Let A be a *totally ordered alphabet* and A^* be the free monoid generated by A. A *rearrangement* of a nonempty word $w = a_1 a_2 \cdots a_m$ is a word $w' = a_{i_1} a_{i_2} \cdots a_{i_m}$, where $i_1 i_2 \cdots i_m$ is a permutation of $1 2 \cdots m$. The set of all the rearrangements of a word w is called a *rearrangement class* (or *abelian class*). Given a subset X of A^* and two integral-valued functions D and E defined on X, the problem is to construct a bijection of X onto itself that maps each word w in X onto a rearrangement w' of w with the subsidiary property that

$$D(w') = E(w). \tag{10.1.1}$$

In most cases the set X is a union of rearrangement classes. It then suffices to give the construction on each such a class. From (10.1.1) it follows that for each integer k and each rearrangement class Y contained in X we have

$$\mathrm{Card}\{w \in Y \,|\, D(w) = k\} = \mathrm{Card}\{w \in Y \,|\, E(w) = k\}. \tag{10.1.2}$$

In probabilistic language this simply means that the statistics D and E are *identically distributed* on Y.

One of the reasons for constructing such bijections is to discover further refinement properties of the distributions of the statistics involved (see, for example, Section 10.7 and Problem 10.6.3). As will be seen, the constructions of those bijections, also called *rearrangements*, make use of the classical techniques described in this book (factorizations, cyclic shifts, and so on).

The first example of such a rearrangement is the "first fundamental transformation" (see Section 10.2). It is a bijection of the permutation group of the set of n elements onto itself. Its construction is based upon the fact that each permutation of $12\cdots n$ can be expressed either as a word $a_1 a_2 \cdots a_n$, or as a product of disjoint cycles. It is worth noting that such a simple construction already gives nontrivial results about the distributions of several statistics defined on the permutation group.

The first fundamental transformation is further extended to each arbitrary set of words (having repetitions). There is some algebraic work to do in order to achieve that extension. In particular a substitute for the notion of cycle, which the first fundamental transformation was based upon, has to be found. The algebraic structures to be introduced are first the *flow monoid* (that is, a quotient monoid of $(A \times A)^*$ derived by a set of commutation rules), then a submonoid called the *circuit monoid*. This will be discussed in Sections 10.3 and 10.4. The first fundamental transformation is then described in Section 10.5. Finally Sections 10.6 and 10.7 give a description of the second fundamental transformation, on the one hand, and of the Sparre-Andersen equivalence principle, on the other hand.

10.2. The First Fundamental Transformation

Let n be a (strictly) positive integer and w be a permutation of the set $[n] = \{1, 2, \ldots, n\}$. For each $i = 1, 2, \ldots, n$ let a_i be the image of i under w. The word $a_1 a_2 \cdots a_n$ will be referred to as the *standard* word associated to w and also denoted by w. Assuming that the alphabet A contains the set $\mathbb{N}$ of the natural numbers, the permutation group $\mathfrak{S}_n$ may be regarded as a subset of A^*. Before constructing the first fundamental transformation for each set $\mathfrak{S}_n$ we mention a few notations valid for arbitrary words, not necessarily standard.

Let $w = a_1 a_2 \cdots a_m$ be a nonempty word. Its *first letter* a_1 is denoted by Fw. Rewriting its m letters in nondecreasing order, we obtain its *nondecreasing rearrangement* denoted by $\bar{w} = \bar{a}_1 \bar{a}_2 \cdots \bar{a}_m$ ($\bar{a}_1 \leqslant \bar{a}_2 \leqslant \cdots \leqslant \bar{a}_m$). If w is an element of the permutation group $\mathfrak{S}_m$, its nondecreasing rearrangement is $12 \cdots m$. If the letters of w are not distinct, containing (say) m_1 letters $1, m_2$ letters $2, \ldots, m_n$ letters n, then

$$\bar{w} = 1^{m_1} 2^{m_2} \cdots n^{m_n}. \tag{10.2.1}$$

Let (a, b) be an ordered pair of integers. Denote by $\nu_{a,b}(w)$ (resp. $\xi_{a,b}(w)$) the number of integers i such that $1 \leqslant i \leqslant m - 1$ and $\bar{a}_i = a, a_i = b$ (resp. $1 \leqslant i \leqslant m - 1$ and $a_i = b, a_{i+1} = a$). Clearly $\nu_{a,b}(w)$ and $\xi_{a,b}(w)$ can only be equal to 0 or 1 if w is standard. The numbers

$$E(w) = \sum_{a<b} \nu_{a,b}(w) \qquad D(w) = \sum_{a<b} \xi_{a,b}(w) \qquad (10.2.2)$$

are frequently referred to as being the number of *exceedances* and number of *descents* of w, respectively. Each word w is said to be *initially dominated* if $a_1 > a_i$ holds for all i with $2 \leqslant i \leqslant n$. Finally, an *increasing factorization* of w is a sequence $(w_1, w_2, \ldots, w_p)$ of initially dominated words with the property that

$$w = w_1 w_2 \cdots w_p$$

and

$$Fw_1 \leqslant Fw_2 \leqslant \cdots \leqslant Fw_p.$$

For instance, the words $w = 563182947$ and $w' = 311264622665175$ admit the increasing factorizations $(5, 631, 82, 947)$ and $(3112, 64, 622, 6, 651, 75)$, respectively.

As shown in the following lemma, increasing factorizations are in fact factorizations of the free monoid in the sense of Chapter 5 (see Problem 5.4.2).

LEMMA 10.2.1. *Every word* $w = a_1 a_2 \cdots a_n$ *admits one and only one increasing factorization.*

Proof. Say that the letter a_i is outstanding in w if $i = 1$ or $2 \leqslant i \leqslant n$ and $a_j \leqslant a_i$ for all $j \leqslant i - 1$. When cutting the word w just before each outstanding letter we clearly obtain an increasing factorization. It remains to be shown that it is the only one.

Suppose that there are two such factorizations, say $(v_1, v_2, \ldots, v_r)$ and $(w_1, w_2, \ldots, w_s)$. Let j be the smallest index such that $v_j \neq w_j$. We can assume that v_j is shorter that w_j, so that $w_j = v_j u$ for some nonempty word u and $Fu = Fv_{j+1}$. As w_j is initially dominated, we have $Fw_j > Fu = Fv_{j+1}$. On the other hand, as $(v_1, v_2, \ldots, v_p)$ is an increasing factorization, we get $Fw_j = Fv_j \leqslant Fv_{j+1}$, leading to a contradiction. Thus the factorization is unique. ∎

The construction of the first fundamental transformation (in the permutation case) goes as follows: First, let τ be a cyclic permutation of a finite set $B = \{b_1, b_2, \ldots, b_m\}$ of m integers. Then, define $q(\tau)$ as the following word of length m:

$$q(\tau) = \tau^m (\max B) \tau^{m-1} (\max B) \cdots \tau (\max B).$$

As τ is cyclic, we have $\tau^m(\max B) = \max B$. Furthermore, $q(\tau)$ is a rearrangement of the m elements of B in some order. Clearly, q is a bijection of the set of cyclic permutations of B onto the set of initially dominated rearrangements of the word $b_1 b_2 \cdots b_m$.

Now let $w = a_1 a_2 \cdots a_n$ be the standard word associated to a permutation of the set $[n]$. If the permutation has r orbits $B_1, B_2, \ldots, B_r$, we can assume that those orbits are numbered in such a way that

$$\max B_1 < \max B_2 < \cdots < \max B_r. \tag{10.2.3}$$

Let $\tau_1, \tau_2, \ldots, \tau_r$ be the restrictions of w to $B_1, B_2, \ldots, B_r$, respectively. As they are all cyclic permutations, we can form the words $q(\tau_1), q(\tau_2), \ldots, q(\tau_r)$. We then let $\hat{w}$ be equal to the juxtaposition product,

$$\hat{w} = q(\tau_1) q(\tau_2) \cdots q(\tau_r).$$

The sequence $(q(\tau_1), q(\tau_2), \ldots, q(\tau_r))$ is precisely the increasing factorization of $\hat{w}$. The mapping that associates $\hat{w}$ to w is a bijection, since there corresponds to w one and only one sequence of cyclic permutations of sets $B_1, B_2, \ldots, B_r$, with union $\{1, 2, \ldots, n\}$ such that (10.2.3) holds. To such a sequence $(\tau_1, \tau_2, \ldots, \tau_r)$ there corresponds next one and only one sequence $(w_1, w_2, \ldots, w_r)$ of initially dominated words such that $F w_1 < F w_2 < \cdots < F w_r$ and $w_1 w_2 \cdots w_r$ be a rearrangement of $12 \cdots n$. From Lemma 10.2.1 there finally corresponds to $(w_1, w_2, \ldots, w_r)$ one and only one permutation $\hat{w}$ admitting $(w_1, w_2, \ldots, w_r)$ as its increasing factorization.

Example 10.2.2. Consider the permutation

$$w = \begin{pmatrix} 1 & 2 & 3 & 4 & 5 & 6 & 7 & 8 & 9 \\ 3 & 8 & 6 & 9 & 5 & 1 & 4 & 2 & 7 \end{pmatrix}.$$

The orbits written in increasing order according to their maxima are

$$B_1 = \{5\}, \quad B_2 = \{1, 3, 6\}, \quad B_3 = \{2, 8\}, \quad B_4 = \{4, 7, 9\}.$$

Let τ_j be the restriction of w to B_j $(1 \leq j \leq 4)$. Then

$$q(\tau_1) = 5;$$
$$q(\tau_2) = \tau_2^3(6) \tau_2^2(6) \tau_2(6) = 631;$$
$$q(\tau_3) = \tau_3^2(8) \tau_3(8) = 82;$$
$$q(\tau_4) = \tau_4^3(9) \tau_4^2(9) \tau_4(9) = 947.$$

Hence

$$\hat{w} = 563182947.$$

Going back to the general case the construction of the inverse bijection is made as follows: Start with a permutation v and consider the increasing factorization, say $(w_1, w_2, \ldots, w_r)$ of v. The product (in the group-theoretic sense) of the disjoint cycles $q^{-1}(w_1) q^{-1}(w_2) \cdots q^{-1}(w_r)$ is a permutation of the set $\{1, 2, \ldots, n\}$. There corresponds to it a unique standard word w. Then $\hat{w} = v$.

Working again with the foregoing example, with $v = \hat{w}$, note that the increasing factorization of v reads $(5, 631, 82, 947)$. We can then form the product of the disjoint cycles:

$$\begin{pmatrix} 5 \\ 5 \end{pmatrix} \begin{pmatrix} 3 & 1 & 6 \\ 6 & 3 & 1 \end{pmatrix} \begin{pmatrix} 2 & 8 \\ 8 & 2 \end{pmatrix} \begin{pmatrix} 4 & 7 & 9 \\ 9 & 4 & 7 \end{pmatrix},$$

Erasing the parenthesis and rearranging the columns in such a way that the top row is in increasing order we obtain the permutation

$$\begin{pmatrix} 1 & 2 & 3 & 4 & 5 & 6 & 7 & 8 & 9 \\ 3 & 8 & 6 & 9 & 5 & 1 & 4 & 2 & 7 \end{pmatrix}.$$

The standard word w such that $\hat{w} = v$ is simply the bottom row of the latter matrix.

The first fundamental transformation is now used to prove the following combinatorial theorem, which essentially says that the number of exceedances E and the number of descents D (see Eqs. (10.2.2)) are identically distributed on each permutations group $\mathfrak{S}_n$.

THEOREM 10.2.3. *For each pair of integers (a, b) with $a < b$ and each standard word w we have*

$$\nu_{a,b}(w) = \xi_{a,b}(\hat{w}). \tag{10.2.3}$$

In particular

$$E(w) = D(\hat{w}).$$

Proof. Let $w = a_1 a_2 \cdots a_n$. If $\nu_{a,b}(w) = 1$, then $a = i < a_i = b$ for some i. But i and a_i belong to the same orbit, say I_j. Let τ_j be the restriction of w to I_j. Then, the dominated word $q(\tau_j)$ contains the factor $a_i i$ — that is, $b a$ — and so $\xi_{a,b}(\hat{w}) = 1$. Conversely, let $\hat{w} = b_1 b_2 \cdots b_n$. If $\xi_{a,b}(\hat{w}) = 1$, there is one and only one factor $b_i b_{i+1}$ of $\hat{w}$ that is equal to $b a$. Let $(w_1, w_2, \ldots, w_p)$ be the increasing sequence of $\hat{w}$. As $b_i > b_{i+1}$, the letter b_i cannot be the last letter of a word w_j and b_{j+1} be the first letter of w_{j+1}. Hence $b_i b_{i+1} = b\,a$ is a proper factor of some word w_j. With $\tau_j = q^{-1}(w_j)$ we then have $b = \tau_j(a)$. Thus b is the image of a under w; that is, $\nu_{a,b}(w) = 1$. The second part of the theorem is an immediate consequence of the first part and definition (10.2.2.).   ∎

It follows from Theorem 10.2.3 that for each integer k we have

$$\text{Card}\{w \in \mathfrak{S}_n \,|\, E(w) = k\} = \text{Card}\{w \in \mathfrak{S}_n \,|\, D(w) = k\}$$

Their common value is the *Eulerian number* denoted by $A_{n,k}$. (See Problems 10.2.1–10.2.3.)

10.3. The Flow Monoid

Denote by $M(A)$ the free monoid generated by the cartesian product $A \times A$. It will be convenient to consider the elements of $M(A)$ as two-row matrices $W = \begin{pmatrix} w' \\ w' \end{pmatrix}$ with w and w' two words of A^* of the same length. Two elements W_1 and W_2 of $M(A)$ are said to be *adjacent* if there exist U and V in $M(A)$ and two one-column matrices $\begin{pmatrix} a' \\ a \end{pmatrix}, \begin{pmatrix} b' \\ b \end{pmatrix}$ with a, a', b, b' in A, having the property that

$$a' \neq b' \tag{10.3.1}$$

and

$$W_1 = U \begin{pmatrix} a' \\ a \end{pmatrix} \begin{pmatrix} b' \\ b \end{pmatrix} V, \qquad W_2 = U \begin{pmatrix} b' \\ b \end{pmatrix} \begin{pmatrix} a' \\ a \end{pmatrix} V. \tag{10.3.2}$$

Notice condition (10.3.1). The commutation rule refers only to the *top* rows of the matrices. Moreover, two adjacent matrices differ by two adjacent columns whose top elements are *distinct*. Next two elements W_1 and W_2 are said to be *equivalent* if they are equal or if there exist an integer $p \geqslant 1$ and a sequence of elements $V_0, V_1, \ldots, V_p$ of $M(A)$ such that $W_1 = V_0, W_2 = V_p$ and V_{i-1} and V_i are adjacent for $1 \leqslant i \leqslant p$. The equivalence relation just defined is compatible with the juxtaposition product in $M(A)$. The quotient monoid of $M(A)$ derived by this equivalence relation is called the *flow monoid* and denoted by $F(A)$. Its elements are called *flows*. The equivalence class of an element $W = \begin{pmatrix} w' \\ w \end{pmatrix}$ will be denoted by $[W] = \begin{bmatrix} w' \\ w \end{bmatrix}$. The map $\begin{pmatrix} a \\ b \end{pmatrix} \mapsto \begin{bmatrix} a \\ b \end{bmatrix}$ is an injection of $A \times A$ into $F(A)$, and $F(A)$ is generated by the set of all $\begin{bmatrix} a \\ b \end{bmatrix}$ with a, b in A. If v is a word of length m (where $m \geqslant 1$) and a an element of A, the flow $\begin{bmatrix} a^m \\ v \end{bmatrix}$ has a single representative, namely $\begin{pmatrix} a^m \\ v \end{pmatrix}$.

In the next lemma is determined an invariant of a flow. Let $W = \begin{pmatrix} w' \\ w \end{pmatrix}$ be the two-row matrix with $w = a_1 a_2 \cdots a_m$ and $w' = a'_1 a'_2 \cdots a'_m$. For each a in A let $(i_1, i_2, \ldots, i_p)$ be the increasing sequence of integers i such that $1 \le i \le m$ and $a'_i = a$. Then W^a will denote the subword $a_{i_1} a_{i_2} \cdots a_{i_p}$ of w.

For instance, for

$$W = \begin{pmatrix} w' \\ w \end{pmatrix} = \begin{pmatrix} 3 & 1 & 1 & 2 & 4 & 1 & 5 \\ 2 & 3 & 1 & 2 & 2 & 4 & 1 \end{pmatrix},$$

we obtain

$$W^1 = 314.$$

Of course W^a is the empty word if a does not occur in w'.

LEMMA 10.3.1. *Let* $W_1 = \begin{pmatrix} w'_1 \\ w_1 \end{pmatrix}$ *and* $W_2 = \begin{pmatrix} w'_2 \\ w_2 \end{pmatrix}$ *be two equivalent elements of $M(A)$. Then*

(i) $W_1^a = W_2^a$ *holds for every a in A;*
(ii) *The word w_2 (resp. w'_2) is a rearrangement of w_1 (resp. w'_1).*

Proof. Properties (i) and (ii) trivially hold when W_1 and W_2 are adjacent. Hence, they are also true for any two equivalent elements of $M(A)$. ∎

THEOREM 10.3.2. (i) *Each nonempty flow f has a unique factorization of the form*

$$f = \begin{bmatrix} a_1^{m_1} \\ v_1 \end{bmatrix} \begin{bmatrix} a_2^{m_2} \\ v_2 \end{bmatrix} \cdots \begin{bmatrix} a_n^{m_n} \\ v_n \end{bmatrix}, \qquad (10.3.3)$$

with $a_1, a_2, \ldots, a_n$ in A satisfying $a_1 < a_2 < \cdots < a_n$ and $m_1 \ge 1, m_2 \ge 1, \ldots, m_n \ge 1$.

(ii) *If $W = \begin{pmatrix} w' \\ w \end{pmatrix}$ is any representative of f, then the word $a_1^{m_1} a_2^{m_2} \cdots a_n^{m_n}$ is the nondecreasing rearrangement of w'.*

(iii) *Finally*

$$W^{a_i} = v_i \quad \text{for each} \quad i = 1, 2, \ldots, n. \qquad (10.3.4)$$

Proof. Clearly (ii) and (iii) are consequences of (i) and the previous lemma. Let us establish (i).

As each flow of the form $\begin{bmatrix} a^m \\ v \end{bmatrix}$ with a in A and v in A^* has a single representative $\begin{pmatrix} a^m \\ v \end{pmatrix}$, it suffices to prove that each nonempty element of

$M(A)$ is equivalent to exactly one product of the form

$$\begin{pmatrix} a_1^{m_1} \\ v_1 \end{pmatrix}\begin{pmatrix} a_2^{m_2} \\ v_2 \end{pmatrix}\cdots\begin{pmatrix} a_n^{m_n} \\ v_n \end{pmatrix},$$

that is, to a matrix $\begin{pmatrix} v' \\ v \end{pmatrix}$ with v' nondecreasing.

Existence. Let $W = \begin{pmatrix} w' \\ w \end{pmatrix}$ be an element of $M(A)$ and assume that the nondecreasing rearrangement of w' is $a_1^{m_1} a_2^{m_2} \cdots a_n^{m_n}$. There is nothing to prove when W has length 1. Assume now that the length of W is at least 2. Denote by w_2' the longest right factor of w' having no letter equal to a_n. Then $w' = w_1' a_n w_2'$ for some word w_1'. Also

$$W = \begin{pmatrix} w_1' \\ w_1 \end{pmatrix}\begin{pmatrix} a_n \\ b \end{pmatrix}\begin{pmatrix} w_2' \\ w_2 \end{pmatrix}$$

for some words w_1, w_2, and some letter b. But

$$\begin{pmatrix} w_1' & w_2' & a_n \\ w_1 & w_2 & b \end{pmatrix}$$

is equivalent to

$$W = \begin{pmatrix} w_1' & a_n & w_2' \\ w_1 & b & w_2 \end{pmatrix}.$$

By induction the matrix

$$\begin{pmatrix} w_1' & w_2' \\ w_1 & w_2 \end{pmatrix}$$

is equivalent to an element $\begin{pmatrix} u' \\ u \end{pmatrix}$ with u' nondecreasing. As u' is a rearrangement of $w_1' w_2'$ the word $u' a_n$ is nondecreasing. Thus the element $\begin{pmatrix} u' & a_n \\ u & b \end{pmatrix}$ has the desired property. Moreover, it is equivalent to

$$\begin{pmatrix} w_1' & w_2' & a_n \\ w_1 & w_2 & b \end{pmatrix}$$

and, a fortiori, to $\begin{pmatrix} w' \\ w \end{pmatrix}$ for the equivalence relation is compatible with the product.

Unicity. Let $W^{a_i} = v_i$ for $i = 1, 2, \ldots, n$. Further, let

$$\begin{pmatrix} v' \\ v \end{pmatrix} = \begin{pmatrix} a_1^{m_1} & a_2^{m_2} & \cdots & a_n^{m_n} \\ v_1 & v_2 & \cdots & v_n \end{pmatrix}.$$

We show that each element $T = \begin{pmatrix} t' \\ t \end{pmatrix}$ equivalent to $\begin{pmatrix} w' \\ w \end{pmatrix}$, with t' nonde-

creasing, is necessarily equal to $\begin{pmatrix} v' \\ v \end{pmatrix}$. It follows from Lemma 10.3.1 (ii) that

t' is equal to $a_1^{m_1} a_2^{m_2} \cdots a_n^{m_n}$. Hence $\begin{pmatrix} t' \\ t \end{pmatrix}$ is equal to the product

$$\begin{pmatrix} a_1^{m_1} \\ t_1 \end{pmatrix} \begin{pmatrix} a_2^{m_2} \\ t_2 \end{pmatrix} \cdots \begin{pmatrix} a_n^{m_n} \\ t_n \end{pmatrix}$$

for some words $t_1, t_2, \ldots, t_n$. But $T^{a_i} = t_i$ (where $1 \leqslant i \leqslant n$). As T is equivalent to W, Lemma 10.3.1 (i) implies that $W^{a_i} = T^{a_i}$; that is, $v_i = t_i$ for each $i = 1, 2, \ldots, n$. Hence $t = t_1 t_2 \cdots t_n = v_1 v_2 \cdots v_n$. ∎

COROLLARY 10.3.3. *The cancellation law holds in the flow monoid $F(A)$. In other words, for any flows f, f', f'' the equality $ff'' = f'f''$ (resp. $f''f = f''f'$) implies $f = f'$.*

Proof. Let f be a flow and a, b two letters. Consider the equation in g

$$f = g \begin{bmatrix} a \\ b \end{bmatrix}. \tag{10.3.5}$$

Using the factorization of f given in Theorem 10.3.2 we have

$$\begin{bmatrix} a_1^{m_1} \\ v_1 \end{bmatrix} \begin{bmatrix} a_2^{m_2} \\ v_2 \end{bmatrix} \cdots \begin{bmatrix} a_n^{m_n} \\ v_n \end{bmatrix} = g \begin{bmatrix} a \\ b \end{bmatrix}.$$

But this equation has a solution only if a is equal to some a_i $(1 \leqslant i \leqslant n)$. If $a = a_i$ (say), then $v_i = wb$ for some word w. Let

$$h = \begin{bmatrix} a_1^{m_1} \\ v_1 \end{bmatrix} \cdots \begin{bmatrix} a_i^{m_i - 1} \\ w \end{bmatrix} \cdots \begin{bmatrix} a_n^{m_n} \\ v_n \end{bmatrix}.$$

Then

$$f = h \begin{bmatrix} a \\ b \end{bmatrix},$$

so that h is a solution of (10.3.5). Any other solution is of the form

$$g = \begin{bmatrix} a_1^{m_1} \\ u_1 \end{bmatrix} \cdots \begin{bmatrix} a_i^{m_i - 1} \\ u_i \end{bmatrix} \cdots \begin{bmatrix} a_n^{m_n} \\ u_n \end{bmatrix}.$$

As

$$h\begin{bmatrix} a \\ b \end{bmatrix} = g\begin{bmatrix} a \\ b \end{bmatrix}$$

we conclude from Theorem 10.3.2 that necessarily $u_1 = v_1, \ldots, u_i b = wb, \ldots, u_n = v_n$. Therefore (10.3.5) has at most one solution.

Consider now three flows f, f', f''. As

$$f\begin{bmatrix} a \\ b \end{bmatrix} = f'\begin{bmatrix} a \\ b \end{bmatrix}$$

implies $f = f'$, we have by induction on the length of f'' that $ff'' = f'f''$ implies $f = f'$. The "resp." part is proved in the same manner. ∎

10.4. The Circuit Monoid

Note that in each element $\begin{pmatrix} w' \\ w \end{pmatrix}$ of $M(A)$ the word w' is not necessarily a rearrangement of w (although it is of the same length). When it is a rearrangement, the equivalence class $\begin{bmatrix} w' \\ w \end{bmatrix}$ is called a *circuit*. Clearly, the set $C(A)$ of all circuits form a submonoid of $F(A)$, called the *circuit monoid*. It follows from Theorem 10.3.2 that each circuit c has one and only one representative in the form $\begin{bmatrix} \bar{v} \\ v \end{bmatrix}$ with v a word and $\bar{v}$ the nondecreasing rearrangement of v. Let

$$\bar{c} = \bar{v} \qquad\qquad (10.4.1)$$

$$\Pi(c) = v. \qquad\qquad (10.4.2)$$

Conversely, to each word v of A^* there corresponds one and only one two-row matrix of the form $\begin{bmatrix} \bar{v} \\ v \end{bmatrix}$ with $\bar{v}$ the nondecreasing rearrangement of v. Define the *circuit* $\Gamma(v)$ by

$$\Gamma(v) = \begin{bmatrix} \bar{v} \\ v \end{bmatrix}. \qquad\qquad (10.4.3)$$

We then have $\Gamma\Pi(c) = c$ and $\Pi\Gamma(v) = v$ for each c in $C(A)$ and v in A^*. Thus the two maps

$$\Pi: C(A) \to A^* \quad \text{and} \quad \Gamma: A^* \to C(A) \qquad\qquad (10.4.4)$$

are *bijective* and *inverses* of each other. Moreover

$$\overline{\Pi(c)} = \bar{c} \quad \text{and} \quad \overline{\Gamma(v)} = \bar{v}, \qquad\qquad (10.4.5)$$

denoting again by $\bar{v}$ the nondecreasing rearrangement of v.

The definition of Γ (given in (10.4.3)) is straightforward. As for obtaining $\Pi(c)$ (whose definition is shown in (10.4.2)) we can proceed as follows: Take any representative $W = \begin{pmatrix} w' \\ w \end{pmatrix}$ of c and let $a_1 < a_2 < \cdots < a_n$ be the distinct elements of A that occur in w' (or w). Then $\Pi(c)$ is the word $v_1 v_2 \cdots v_n$ with

$$W^{a_i} = v_i \; (i = 1, 2, \ldots, n). \tag{10.4.6}$$

The final step is to define another bijection $\Delta: A^* \to C(A)$, and the fundamental transformation $w \mapsto \hat{w}$ will be the functional product $\Delta^{-1} \circ \Gamma$. For each word $w = a_1 a_2 \cdots a_m$ denote by δw the *cyclic shift*

$$\delta w = a_2 a_3 \cdots a_m a_1.$$

Remember that a word is said to be (initially) dominated if its first letter is (strictly) greater than all its other letters. In the same manner, a circuit c will be said to be *dominated* if

$$c = \begin{bmatrix} \delta w \\ w \end{bmatrix} \tag{10.4.7}$$

with w dominated. Clearly, for each circuit c there is at most one dominated word w such that (10.4.7) holds. When w is dominated, we will denote by $\gamma(w)$ the circuit

$$\gamma(w) = \begin{bmatrix} \delta w \\ w \end{bmatrix}. \tag{10.4.8}$$

Thus γ is a bijection of the set of all dominated words onto the set of dominated circuits. In (10.4.7) the first letter of the dominated word w, previously denoted by Fw, will also be written Fc.

By definition a *dominated circuit factorization* of a circuit c is a sequence $(d_1, d_2, \ldots, d_r)$ of dominated circuits with the property that

$$c = d_1 d_2 \cdots d_r$$

and

$$Fd_1 \leqslant Fd_2 \leqslant \cdots \leqslant Fd_r. \tag{10.4.9}$$

THEOREM 10.4.1. *Each nonempty circuit admits exactly one dominated circuit factorization.*

The proof of Theorem 10.4.1 actually gives the construction of the factorization. The dominated circuits are to be sorted one by one out of the initial circuit. Let us first prove the following lemma:

LEMMA 10.4.2. *Let v be a nonempty word and $\bar{v}$ be its nondecreasing rearrangement. If there exists an integer $i \geq 1$, a sequence $(a_0'', a_1'', \ldots, a_i'')$ of letters and an element $\begin{pmatrix} v_i' \\ v_i \end{pmatrix}$ of $M(A)$ with the properties that*

(i) *a_0'' is different from each of the letters $a_1'', \ldots, a_i''$;*
(ii) *$\begin{pmatrix} \bar{v} \\ v \end{pmatrix}$ is equivalent to*

$$\begin{pmatrix} v_i' \\ v_i \end{pmatrix} \begin{pmatrix} a_{i-1}'' & a_{i-2}'' & \cdots & a_1'' & a_0'' \\ a_i'' & a_{i-1}'' & \cdots & a_2'' & a_1'' \end{pmatrix},$$

then a letter a_{i+1}'' and a matrix $\begin{pmatrix} v_{i+1}' \\ v_{i+1} \end{pmatrix}$ of $M(A)$ can be found so that condition (ii) *holds when i is replaced by $i+1$.*

Proof. Define a_{i+1}'' as the bottom element of the rightmost one-column submatrix of $\begin{pmatrix} v_i' \\ v_i \end{pmatrix}$ whose top element is equal to a_i''. This definition makes sense, for conditions (i) and (ii) imply that v_i' contains a letter equal to a_i''. Hence, the following factorization holds:

$$\begin{pmatrix} v_i' \\ v_i \end{pmatrix} = \begin{pmatrix} u_1' \\ u_1 \end{pmatrix} \begin{pmatrix} a_i'' \\ a_{i+1}'' \end{pmatrix} \begin{pmatrix} u_2' \\ u_2 \end{pmatrix}.$$

with no letter equal to a_i'' in u_2'. Then put

$$\begin{pmatrix} v_{i+1}' \\ v_{i+1} \end{pmatrix} = \begin{pmatrix} u_1' & u_2' \\ u_1 & u_2 \end{pmatrix}.$$

The following two elements of $M(A)$:

$$\begin{pmatrix} v_i' \\ v_i \end{pmatrix} \quad \text{and} \quad \begin{pmatrix} v_{i+1}' \\ v_{i+1} \end{pmatrix} \begin{pmatrix} a_i'' \\ a_{i+1}'' \end{pmatrix}$$

are equivalent. Thus condition (i) also holds when i is replaced by $i+1$. ∎

We are now ready to complete the proof of Theorem 10.4.1. Let $v = b_1 b_2 \cdots b_m$ be a word with a nondecreasing rearrangement given by $\bar{v} = a_1' a_2' \cdots a_m' = a_1^{m_1} a_2^{m_2} \cdots a_n^{m_n}$ (where $a_1 < a_2 < \cdots < a_n$; $m_1 \geq 1, m_2 \geq 1, \ldots, m_n \geq 1$). Consider the circuit

$$z = \Gamma(v) = \begin{bmatrix} \bar{v} \\ v \end{bmatrix} = \begin{bmatrix} a_1' & a_2' & \cdots & a_m' \\ b_1 & b_2 & \cdots & b_m \end{bmatrix}.$$

If v is of length one, then

$$c = \begin{bmatrix} \bar{v} \\ v \end{bmatrix} = \begin{bmatrix} \delta v \\ v \end{bmatrix}$$

and the theorem is proved. Assume $m \geqslant 2$. If $b_m = a'_m \ (= a_n)$, let

$$c' = \begin{bmatrix} a'_1 & a'_2 & \cdots & a'_{m-1} \\ b_1 & b_2 & \cdots & b_{m-1} \end{bmatrix} \quad \text{and} \quad d = \begin{bmatrix} a_n \\ a_n \end{bmatrix}.$$

Then

$$c = c'd.$$

By induction on m, the circuit c' admits the unique dominated circuit factorization

$$c' = d_1 d_2 \cdots d_r.$$

As d is trivially dominated and a_n is the maximum letter of v, we also have

$$Fd_1 \leqslant Fd_2 \leqslant \cdots \leqslant Fd_r \leqslant Fd.$$

Therefore c has the following dominated circuit factorization

$$c = d_1 d_2 \cdots d_r d.$$

If $b_m \neq a'_m$, let $a''_1 = b_m$, $a''_0 = a'_m$, $v'_1 = a'_1 a'_2 \cdots a'_{m-1}$, and $v_1 = b_1 b_2 \cdots b_{m-1}$. Then conditions (i) and (ii) of Lemma 10.4.2 both hold for $i = 1$. By applying Lemma 10.4.2 inductively we can form a sequence $(a''_0, a''_1, \dots)$ of letters. Let $i + 1$ be the first integer for which Lemma 10.4.2 does not apply. Such an integer exists since the sequence is necessarily finite. We then have $a''_{i+1} = a''_0$, but still a''_0 different from each of the letters $a''_1, \dots, a''_i$. If $i + 1 \leqslant m - 1$, let

$$c' = \begin{bmatrix} v'_{i+1} \\ v_i \end{bmatrix} \quad \text{and} \quad d = \begin{bmatrix} a''_i & a''_{i-1} & \cdots & a''_1 & a''_0 \\ a''_0 & a''_i & \cdots & a''_2 & a''_1 \end{bmatrix}.$$

Then d is dominated (by the maximum letter $a''_0 = a_n$). If $d_1 d_2 \cdots d_r$ is the dominated circuit factorization of c', then we conclude, as before, that

$$d_1 d_2 \cdots d_r d$$

is a dominated circuit factorization of c.

It remains for us to prove the unicity of the factorization. Let $c_1 c_2 \cdots c_q$ and $d_1 d_2 \cdots d_r$ be two dominated circuit factorizations of a circuit $c = \Gamma(v)$.

Let $v_q = b'_1 b'_2 \cdots b'_s$ and $w_r = b''_1 b''_2 \cdots b''_t$ be two dominated words defined by $\gamma(v_q) = c_q$ and $\gamma(w_r) = d_r$ (see (10.4.8)). Thus

$$c_q = \begin{bmatrix} \delta v_q \\ v_q \end{bmatrix} = \begin{bmatrix} b'_2 & \cdots & b'_s & b'_1 \\ b'_1 & \cdots & b'_{s-1} & b'_s \end{bmatrix}$$

and

$$d_r = \begin{bmatrix} \delta w_r \\ w_r \end{bmatrix} = \begin{bmatrix} b''_2 & \cdots & b''_t & b''_1 \\ b''_1 & \cdots & b''_{t-1} & b''_t \end{bmatrix}.$$

But b'_1 and b''_1 are both equal to the maximum letter of v. Therefore $b'_1 = b''_1$. Assume $s \leqslant t$. By induction $b'_{s-1} = b''_{t-1}, \ldots, b'_1 = b''_{t-s+1}$. But $b'_1 = b''_1$. As w_r is dominated, the equation $b''_1 = b''_{t-s+1}$ can hold only if $t = s$. Hence $v_q = w_r$ and $c_q = d_r$. As the cancellation law holds in $F(A)$ (see Corollary 10.3.3), we obtain $c_1 c_2 \cdots c_{q-1} = d_1 d_2 \cdots d_{r-1}$. The unicity follows by induction on the length. ∎

10.5. The First Fundamental Transformation for Arbitrary Words

We are now ready to define the second bijection $\Delta: A^* \to C(A)$. Let $(w_1, w_2, \ldots, w_r)$ be the increasing factorization of a word w (see Lemma 10.2.1). Remember that each factor w_i is dominated so that we can form the dominated circuit

$$\gamma(w_i) = \begin{bmatrix} \delta w_i \\ w_i \end{bmatrix}, \quad (i = 1, 2, \ldots, r).$$

Taking their product in $C(A)$ we obtain the circuit

$$\Delta(w) = \gamma(w_1) \gamma(w_2) \cdots \gamma(w_r). \tag{10.5.1}$$

By construction

$$\overline{\Delta(w)} = \overline{w} \tag{10.5.2}$$

(see definition (10.4.1)). On the other hand, γ is a bijection of the set of dominated words onto the set of dominated circuits with the property that $\overline{\gamma(w)} = \overline{w}$ (see (10.4.8)). The map that associates $(w_1, w_2, \ldots, w_r)$ to $(\gamma(w_1), \gamma(w_2), \ldots, \gamma(w_r))$ transforms the increasing factorization of w into the dominated circuit factorization of $\Delta(w)$. It then follows from Lemma 10.2.1 and Theorem 10.4.1 that Δ is *bijective*.

The essential property of Δ is that the adjacent letters ba of w with $a < b$ are transformed into vertical pairs $\begin{bmatrix} a \\ b \end{bmatrix}$ in $\Delta(w)$. We make this definition more precise by introducing a function $\eta_{a,b}$ on circuits as follows: Let

$$c = \begin{bmatrix} a_1' & a_2' & \cdots & a_m' \\ a_1 & a_2 & \cdots & a_m \end{bmatrix} = \begin{bmatrix} a_1' \\ a_1 \end{bmatrix}\begin{bmatrix} a_2' \\ a_2 \end{bmatrix} \cdots \begin{bmatrix} a_m' \\ a_m \end{bmatrix}$$

be a circuit. Then $\eta_{a,b}(c)$ is defined as the number of *vertical occurrences* of $\begin{bmatrix} a \\ b \end{bmatrix}$ in c, that is, the number of integers i with $1 \leqslant i \leqslant m$ and $a_i' = a$, $b_i' = b$. Remember that $\xi_{a,b}(w)$ is the number of two-letter factors of the word w that are equal to ba.

THEOREM 10.5.1. *For each nonempty word w of A^* and each ordered pair (a, b) of letters satisfying $a < b$, we have*

$$\xi_{a,b}(w) = \eta_{a,b}(\Delta(w)). \tag{10.5.3}$$

Proof. Let $(w_1, w_2, \ldots, w_r)$ be the increasing factorization of a word w and $a < b$. First, b cannot be the last letter of w_j, while a is the first letter of the successive factor w_{j+1}. Second, each word $\begin{pmatrix} \delta w_j \\ w_j \end{pmatrix}$ of $M(A)$ cannot end with the letter $\begin{pmatrix} a \\ b \end{pmatrix}$. Hence the horizontal pairs ba and vertical pairs $\begin{pmatrix} a \\ b \end{pmatrix}$ can occur only as shown schematically in the next equation

$$\Delta(w) = \begin{bmatrix} \cdots \\ \cdots \end{bmatrix} \cdots \begin{bmatrix} \cdots a \cdots \\ \cdots ba \cdots \end{bmatrix} \cdots \begin{bmatrix} \cdots \\ \cdots \end{bmatrix}.$$

Thus Eq. (10.5.3) holds.   ■

Denote by Δ^{-1} the inverse of Δ that maps $C(A)$ onto A^*. If w is a word, let

$$\hat{w} = \Delta^{-1}(\Gamma(w)). \tag{10.5.4}$$

THEOREM 10.5.2. *The mapping $w \mapsto \hat{w}$ is a bijection of A^* onto itself having the following properties*:

(i) $\hat{w}$ is a rearrangement of w;
(ii) *for each pair (a, b) with $a < b$ then*

$$\nu_{a,b}(w) = \xi_{a,b}(\hat{w}). \tag{10.5.5}$$

Moreover

$$E(w) = D(\hat{w}). \tag{10.5.6}$$

Proof. As both Γ and Δ^{-1} are bijective, their composition product is also bijective. Property (i) follows from (10.4.5) and (10.5.2). Now let $w = a_1 a_2 \cdots a_m$ be a word and $\bar{w} = a_1' a_2' \cdots a_m'$ be its nondecreasing rearrangement. Remember that $\nu_{a,b}(w)$ is the number of integers i with $a_i' = a$ and $a_i = b$. The relation $\nu_{a,b}(w) = \eta_{a,b}(\Gamma(w))$ for $a < b$ is a trivial consequence of the definitions of $\nu_{a,b}$ and $\eta_{a,b}$. From (10.5.3) we deduce that

$$\xi_{a,b}(\hat{w}) = \eta_{a,b}(\Delta(\hat{w})) = \eta_{a,b}(\Gamma(w)) = \nu_{a,b}(w).$$

As for (10.5.6) it follows from (10.5.5) and definition (10.2.2). ∎

Example 10.5.3. Let us illustrate with an example the construction of the bijection $w \mapsto \hat{w}$ and its inverse. Consider the word

$$w = 31514226672615.$$

Its nondecreasing rearrangement reads

$$\bar{w} = 11122234556667.$$

Then $\Gamma(w)$ (see (10.4.3)) is the circuit

$$\Gamma(w) = \begin{bmatrix} \bar{w} \\ w \end{bmatrix} = \begin{bmatrix} 1 & 1 & 1 & 2 & 2 & 2 & 3 & 4 & 5 & 5 & 6 & 6 & 6 & 7 \\ 3 & 1 & 5 & 1 & 4 & 2 & 2 & 6 & 6 & 7 & 2 & 6 & 1 & 5 \end{bmatrix}.$$

To compute $\Delta^{-1}(\Gamma(w))$ we have to determine the dominated circuit factorization of $\Gamma(w)$ as indicated in Theorem 10.4.1. This is achieved by sorting out the successive dominated circuits from right to left:

$$\begin{aligned}
\Gamma(w) &= \begin{bmatrix} 1 & 1 & 1 & 2 & 2 & 2 & 3 & 4 & 5 & 6 & 6 & 6 \\ 3 & 1 & 5 & 1 & 4 & 2 & 2 & 6 & 6 & 2 & 6 & 1 \end{bmatrix}\begin{bmatrix} 5 & 7 \\ 7 & 5 \end{bmatrix} \\
&= \begin{bmatrix} 1 & 1 & 2 & 2 & 2 & 3 & 4 & 6 & 6 \\ 3 & 1 & 1 & 4 & 2 & 2 & 6 & 2 & 6 \end{bmatrix}\begin{bmatrix} 5 & 1 & 6 \\ 6 & 5 & 1 \end{bmatrix}\begin{bmatrix} 5 & 7 \\ 7 & 5 \end{bmatrix} \\
&= \begin{bmatrix} 1 & 1 & 2 & 2 & 2 & 3 & 4 & 6 \\ 3 & 1 & 1 & 4 & 2 & 2 & 6 & 2 \end{bmatrix}\begin{bmatrix} 6 \\ 6 \end{bmatrix}\begin{bmatrix} 5 & 1 & 6 \\ 6 & 5 & 1 \end{bmatrix}\begin{bmatrix} 5 & 7 \\ 7 & 5 \end{bmatrix} \\
&= \begin{bmatrix} 1 & 1 & 2 & 3 \\ 3 & 1 & 1 & 2 \end{bmatrix}\begin{bmatrix} 4 & 2 & 2 & 6 \\ 6 & 4 & 2 & 2 \end{bmatrix}\begin{bmatrix} 6 \\ 6 \end{bmatrix}\begin{bmatrix} 5 & 1 & 6 \\ 6 & 5 & 1 \end{bmatrix}\begin{bmatrix} 5 & 7 \\ 7 & 5 \end{bmatrix}.
\end{aligned}$$

The word $\hat{w} = \Delta^{-1}(\Gamma(w))$ is then the juxtaposition product of the words occurring in the bottom row in the last product, namely

$$\hat{w} = \overset{\frown}{3112}\overset{\frown}{6422}6\overset{\frown}{651}\overset{\frown}{75}.$$

Conversely, to obtain w from $\hat{w}$—that is, $w = \Gamma^{-1}(\Delta(\hat{w}))$—we first form the increasing factorization of $\hat{w}$

$$(3112, 6422, 6, 651, 75),$$

then define $\Delta(\hat{w})$ (see (10.4.1)), namely

$$\Delta(\hat{w}) = \begin{bmatrix} 1 & 1 & 2 & 3 & 4 & 2 & 2 & 6 & 6 & 5 & 1 & 6 & 5 & 7 \\ 3 & 1 & 1 & 2 & 6 & 4 & 2 & 2 & 6 & 6 & 5 & 1 & 7 & 5 \end{bmatrix}.$$

Then we reshuffle all the columns of $\Delta(\hat{w})$ so that the top row is in increasing order (see (10.4.2)). We find again $\Gamma(w)$, and finally w occurs in the bottom row.

In the example boldface type in $\Gamma(w)$ marks the letters of w that are greater than the corresponding letters of $\bar{w}$ above them. We have the vertical pairs

$$\binom{1}{3}, \binom{1}{5}, \binom{2}{4}, \binom{4}{6}, \binom{5}{6}, \binom{5}{7}.$$

In $\hat{w}$ we have the horizontal pairs

$$31, 64, 42, 65, 51 \quad \text{and} \quad 75.$$

Thus $E(w) = 6 = D(\hat{w})$.

10.6. The Second Fundamental Transformation

Let $w = a_1 a_2 \cdots a_m$ be a word of length $m \geqslant 1$. Its *inversion number*, denoted by INV w, is defined as the number of ordered pairs (i, j) with $1 \leqslant i < j \leqslant m$ and $a_i > a_j$. The *down set* of w, denoted by DOWN w, is defined by

$$\text{DOWN} \, w = \{i \mid 1 \leqslant i \leqslant m - 1, \quad a_i > a_{i+1}\}, \qquad (10.6.1)$$

and the *major index* of w, denoted by MAJ w, as the sum (possibly zero) of the elements in DOWN w.

For instance, with

$$\begin{array}{ccccccccc} & 1 & 2 & 3 & 4 & 5 & 6 & 7 & 8 & 9 \\ w = & 4 & 4 & 2 & 3 & 4 & 1 & 3 & 2 & 3, \end{array}$$

we have INV $w = 20$, DOWN $w = \{2, 5, 7\}$, so that MAJ $w = 2 + 5 + 7 = 14$.

MacMahon (1913, 1915) introduced the function MAJ in the study of ordered partitions. Let X be a rearrangement class of A^* (see Section 10.1) and let

$$I(q) = \sum q^{\text{INV} \, w}, \, M(q) = \sum q^{\text{MAJ} \, w} \, (w \in X) \qquad (10.6.2)$$

be the generating functions for X by number of inversions and major index, respectively. MacMahon obtained (1916) the surprising result that $I(q)$ and

$M(q)$ have the same expression (see Problem 10.10). Schützenberger (private communication 1966) raised the problem of finding a bijection Φ of A^* onto itself with the property that for every word w

(i) $\Phi(w)$ is a rearrangement of w;
(ii) INV $\Phi(w) =$ MAJ w.

The bijection Φ described below was found in Foata (1968) and later referred to as the "second fundamental transformation." When Φ is restricted to the permutation group, it has further interesting properties (see Problem 10.6.3).

The construction of Φ goes as follows: Let a be an element of A and w a nonempty word. If the last letter of w is smaller than or equal to (resp. is greater than) a, the word w clearly admits the unique factorization

$$(v_1 b_1, v_2 b_2, \ldots, v_p b_p)$$

called its *a-factorization* having the following properties:

(i) Each b_i $(1 \leqslant i \leqslant p)$ is a letter satisfying $b_i \leqslant a$ (resp. $b_i > a$);
(ii) Each v_i $(1 \leqslant i \leqslant p)$ is a word that is either empty or has all its letters greater than (resp. smaller than or equal to) a.

Then let

$$\gamma_a(w) = b_1 v_1 b_2 v_2 \cdots b_p v_p.$$

(Note that $w = v_1 b_1 v_2 b_2 \cdots v_p b_p$.) The bijection will be defined by induction on the length of the words as follows:

If w has length 1, let

$$\Phi(w) = w; \tag{10.6.3}$$

If $|w| \geqslant 2$, write $w = va$ with a the last letter of w. By induction determine the word $v' = \gamma_a(\Phi(v))$ and let $\Phi(w)$ be the juxtaposition product

$$\Phi(w) = v'a(= \gamma_a(\Phi(v))a). \tag{10.6.4}$$

Let us describe the effective algorithm for Φ.

ALGORITHM 10.6.1. *Let $w = a_1 a_2 \cdots a_m$ be a word;*

1. *Let $i = 1, w'_1 = a_1$;*
2. *If $i = m$, let $\Phi(w) = w'_i$ and stop; else continue;*

3. *If the last letter of w_i' is smaller than or equal to (resp. greater than) a_{i+1}, split w_i' after each letter smaller than or equal to (resp. greater than) a_{i+1};*
4. *In each compartment of w_i' determined by the splits move the last letter to the beginning; let v' be the word obtained after making those moves; let $w_{i+1}' = v'a_{i+1}$; replace i by $i+1$; go to 2.*

For instance, the image under Φ of the word $w = 442341323$ is obtained as follows.

$$w_1' = 4|$$
$$w_2' = 4|4|$$
$$w_3' = 442|$$
$$w_4' = 2|4|4|3|$$
$$w_5' = 2|4|4|3|4|$$
$$w_6' = 2|443|41|$$
$$w_7' = 23|4|4|14|3|$$
$$w_8' = 3|2|4441|3|2|$$
$$\Phi(w) = w_9' = 321444323.$$

The algorithm can be reversed (see Problem 10.6.1).

THEOREM 10.6.2. *The map Φ is bijective. Furthermore, the image $\Phi(w)$ of each word w is a rearrangement of w. Finally, the following identity holds*

$$\mathrm{INV}\,\Phi(w) = \mathrm{MAJ}\,w. \tag{10.6.5}$$

Proof. The first two statements are easy to prove. As for the last one, let $w = a_1 a_2 \cdots a_m$ and for each a in A let $l_a(w)$ (resp. $r_a(w)$) be the number of subscripts i with $1 \le i \le m$ and $a_i \le a$ (resp. $a_i > a$). Of course, $l_a(w) + r_a(w) = |w|$. Furthermore,

$$\mathrm{INV}\,wa = \mathrm{INV}\,w + r_a(w).$$

Now if the last letter of w is less than or equal to a, we have

$$\mathrm{INV}\,\gamma_a(w) = \mathrm{INV}\,w - r_a(w)$$
$$\mathrm{MAJ}\,wa = \mathrm{MAJ}\,w.$$

When the last letter of w is greater than a, we have this time

$$\mathrm{INV}\,\gamma_a(w) = \mathrm{INV}\,w + l_a(w)$$
$$\mathrm{MAJ}\,wa = \mathrm{MAJ}\,w + l_a(w) + r_a(w).$$

Property (10.6.5) is then a consequence of these five relations together with (10.6.4). First

$$\text{INV}\,\Phi(wa) = \text{INV}\,\gamma_a(\Phi(w))a$$
$$= \text{INV}\,\gamma_a(\Phi(w)) + r_a(\gamma_a(\Phi(w)))$$
$$= \text{INV}\,\gamma_a(\Phi(w)) + r_a(w),$$

since $\gamma_a(\Phi(w))$ is only a rearrangement of w. Then, if the last letter of w is less than or equal to a, we get (by induction):

$$\text{INV}\,\Phi(wa) = \text{INV}\,\gamma_a(\Phi(w)) + r_a(w)$$
$$= (\text{INV}\,\Phi(w) - r_a(w)) + r_a(w)$$
$$= \text{MAJ}\,w$$
$$= \text{MAJ}\,wa.$$

Finally, if the last letter of w is greater than a, we have (by induction)

$$\text{INV}\,\Phi(wa) = \text{INV}\,\gamma_a(\Phi(w)) + r_a(w)$$
$$= \text{INV}\,\Phi(w) + l_a(w) + r_a(w)$$
$$= \text{MAJ}\,w + |w|$$
$$= \text{MAJ}\,wa. \qquad \blacksquare$$

Working with the foregoing example $w = 4\ 4\ 2\ 3\ 4\ 1\ 3\ 2\ 3$ and $\Phi(w) = 3\ 2\ 1\ 4\ 4\ 4\ 3\ 2\ 3$, we obtain

$$\text{MAJ}\,w = 14 = \text{INV}\,\Phi(w).$$

10.7. The Sparre-Andersen Equivalence Principle

The Sparre-Andersen equivalence principle was presented in Proposition 5.2.9. Let us recall that if $w = a_1 a_2 \cdots a_m$ is a word of A^* with A the field of the real numbers, we let

$$\sigma(w) = a_1 + a_2 + \cdots + a_m \qquad (10.7.1)$$

be the *total sum* and $\sigma_0(w) = 0, \sigma_1(w) = \sigma(a_1) = a_1, \sigma_2(w) = \sigma(a_1 a_2) = a_1 + a_2, \ldots, \sigma_m(w) = \sigma(a_1 a_2 \cdots a_m) = a_1 + a_2 + \cdots + a_m$ be the *partial sums* of w. For each $k = 0, 1, \ldots, m$ let $\Pi_{m,k}(w)$ be the number of subscripts i for which

- Either $0 \le i \le k-1$ and $\sigma_i(w) \ge \sigma_k(w)$,
- Or $k+1 \le i \le m$ and $\sigma_i(w) > \sigma_k(w)$. $\qquad (10.7.2)$

Thus $\Pi_{m,k}(w)$ is the *number of partial sums greater than or equal to $\sigma_k(w)$* (with the convention that whenever two partial sums are equal the left-hand one is counted before the right-hand one with $\sigma_k(w)$ itself not being counted.)

For instance, with $m = 8, k = 5$, and $w = 1, -2, 0, 3, -1, 1, -2, 1$, the sequence of partial sums is $(0, 1, -1, -1, 2, 1, 2, 0, 1)$, as graphically represented in Figure 10.1. As $\sigma_k(w) = \sigma_5(w) = 1$, we obtain $\Pi_{8,5}(w) = 3$.

Clearly $\Pi_{m,0}(w)$ is the *number of* (strictly) *positive partial sums* that was denoted by $L(w)$ in Chapter 5. Now the *index of the first maximum* in $(\sigma_0(w), \sigma_1(w), \dots, \sigma_m(w))$ is equal to k—a quantity denoted by $\Pi(w)$ in Chapter 5—if and only if $\Pi_{m,k}(w) = 0$. The Sparre-Andersen equivalence principle expresses the fact that in each rearrangement class X there are for each integer k as many words with $L(w) (= \Pi_{m,0}(w)) = k$ as words w' with $\Pi(w') = k$; that is, $\Pi_{m,k}(w') = 0$.

Later Sparre-Andersen (1962) obtained a further extension of his principle as follows:

THEOREM 10.7.1. *Let j, k be two integers with $0 \leqslant j, k \leqslant m$. Then in each rearrangement class X of words of length m there are as many words w with $\Pi_{m,k}(w) = j$ as words w' with $\Pi_{m,j}(w') = k$.*

Theorem 10.7.1 will be proved by constructing a bijection ρ_k of X onto itself with the property that

$$\Pi_{m,k}(w) = j \Leftrightarrow \Pi_{m,j}(\rho_k(w)) = k. \tag{10.7.3}$$

The following bijection ρ was defined in Chapter 5. If w is the empty word 1, let $\rho(w) = w$, while if a is a letter of A, define by induction on the length m of w

$$\rho(wa) = \rho(w)a \quad \text{if} \quad \sigma_{m+1}(wa) \leqslant 0$$
$$= a\rho(w) \quad \text{otherwise.} \tag{10.7.4}$$

Figure 10.1. Graph associated with a sequence of partial sums.

It was shown in Proposition 5.2.9 that ρ was a bijection of each rearrangement class onto itself and also

$$L(w) = \Pi(\rho(w)),$$

that is,

$$\Pi_{m,0}(w) = j \Leftrightarrow \Pi_{m,j}(\rho(w)) = 0. \tag{10.7.5}$$

Clearly (10.7.5) is a particular case of (10.7.3) obtained for $k = 0$. But how is ρ_k to be defined for the other values of k? This is the purpose of this section.

In Example 5.2.8 it was noted that the two sets

$$R = \{r \in A^* \mid r = uv \Rightarrow \sigma(v) > 0\}$$
$$S = \{s \in A^* \mid s = uv \Rightarrow \sigma(u) \leqslant 0\}$$

were submonoids of A^* and each word w' has a unique factorization

$$w' = rs, \quad r \in R, \quad s \in S. \tag{10.7.6}$$

Moreover, it was shown that the length of r is equal to the index of the first maximum in the sequence of the partial sums of w', a result that can also be expressed, if $|w| = m$, by

$$|r| = j \Leftrightarrow \Pi_{m,j}(w') = 0.$$

It follows from (10.7.5) that if

$$\rho(w) = rs, \quad r \in R, \quad s \in S, \tag{10.7.7}$$

then

$$\Pi_{m,0}(w) = |r|. \tag{10.7.8}$$

CONSTRUCTION OF THE BIJECTION ρ_k. Let k be a fixed integer with $0 \leqslant k \leqslant m$ and $w = a_1 a_2 \cdots a_m$ be a word. To obtain $\rho_k(w)$ calculate successively

1. w_1 and w_2 by

$$w = w_1 w_2, \quad |w_1| = k;$$

2. r_1, r_2, s_1, s_2 by

$\rho(\tilde{w}_1) = r_1 s_1, \rho(w_2) = r_2 s_2$, with $r_1, r_2 \in R$ and $s_1, s_2 \in S$ and ρ defined in (10.7.4);

3. u_1 and u_2 by

$$u_1 = \rho^{-1}(r_2 s_1), \qquad u_2 = \rho^{-1}(r_1 s_2);$$

4. $\rho_k(w) = \widetilde{u}_1 u_2$.

THEOREM 10.7.3. *The mapping ρ_k defined previously maps each rearrangement class onto itself and satisfies property* (10.7.3).

Proof. Clearly ρ_k is a rearrangement. As ρ is bijective and the factorization given in (10.7.6) is unique, the map ρ_k is also bijective. On the other hand, $\Pi_{m,k}(w)$ is also the number of subscripts i for which

- Either $0 \leqslant i \leqslant k-1$ and $\sigma(a_{i+1}a_{i+2}\cdots a_k) \leqslant 0$;
- Or $k+1 \leqslant i \leqslant m$ and $\sigma(a_{k+1}a_{k+2}\cdots a_i) > 0$.

Thus

$$\Pi_{m,k}(w) = \Pi_{k,k}(w_1) + \Pi_{m-k,0}(w_2).$$

As the reverse image $\widetilde{w}_1$ of w_1 has the same total sum as w_1, we deduce that

$$\Pi_{k,k}(w_1) = k - \Pi_{k,0}(\widetilde{w}_1).$$

Therefore

$$\Pi_{m,k}(w) = k - \Pi_{k,0}(\widetilde{w}_1) + \Pi_{m-k,0}(w_2). \qquad (10.7.9)$$

Let $|u_1| = |\widetilde{u}_1| = j$. In the same manner

$$\Pi_{m,j}(\rho_k(w)) = j - \Pi_{j,0}(u_1) + \Pi_{m-j,0}(u_2).$$

On the other hand, (10.7.7) and (10.7.8) applied to $\widetilde{w}_1, w_2, u_1$, and u_2 yield

$$\Pi_{k,0}(\widetilde{w}_1) = |r_1|, \qquad \Pi_{m-k,0}(w_2) = |r_2|,$$
$$\Pi_{j,0}(u_1) = |r_2|, \qquad \Pi_{m-j,0}(u_2) = |r_1|.$$

Hence

$$\Pi_{m,k}(w) = k - |r_1| + |r_2|$$
$$= |s_1| + |r_2| = |u_1| = j,$$

and

$$\Pi_{m,j}(\rho_k(w)) = j - |r_2| + |r_1|$$
$$= |s_1| + |r_1| = |w_1| = k.$$

Thus property (10.7.3) is verified.   ∎

Example 10.7.4. The construction of ρ_k can be illustrated with the example shown in Figure 10.1. There $m = 8$, $k = 5$, $w = 1$, $-2, 0, 3, -1, 1, -2, 1$ and $\Pi_{8,5}(w) = j = 3$. Again consider the four steps of the construction of ρ_k

1. $w_1 = 1, -2, 0, 3, -1$; $w_2 = 1, -2, 1$;
2. $\rho(\widetilde{w}_1) = 1, 0, 3, -1, -2$; $\rho(w_2) = 1, -2, 1$.

As the indices of the first maxima of the partial sums of $\rho(\widetilde{w}_1)$ and $\rho(w_2)$ are equal to 3 and 1, respectively, we have

$$r_1 = 1, 0, 3; \qquad s_1 = -1, -2; \qquad r_2 = 1; \qquad s_2 = -2, 1;$$

3. $u_1 = \rho^{-1}(r_2 s_1) = 1, -1, -2$;
   $u_2 = \rho^{-1}(r_1 s_2) = -2, 1, 3, 0, 1$.
4. $\rho_k(w) = \widetilde{u}_1 u_2 = -2, -1, 1, -2, 1, 3, 0, 1$, which corresponds to the partial sum graph drawn in Figure 10.2.

The number of partial sums of $\rho_k(w)$ that are greater than or equal to $\sigma_j(\rho_k(w)) = \sigma_3(\rho_5(w)) = -2$ is equal to $k = 5$.

Notes

The first fundamental transformation for permutations is already implicit in Riordan (1958, Chapter 8). It is an essential tool in the study of Eulerian polynomials, as shown in Foata and Schützenberger (1970). The extension of the first fundamental transformation to arbitrary words was obtained by Foata (1965). Then Cartier and Foata (1969) derived a convenient set-up to describe it first by introducing the monoids subject to commutation rules, second by developing the study of the flow and circuit monoids. Lallement (1977) took up again this study in one chapter of his book. The circuit monoid was also used by Foata (1979, 1980), in particular to derive a

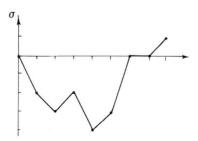

Figure 10.2. Graph associated with a sequence of partial sums.

noncommutative version of the matrix inversion formula. Möbius inversion identities can be obtained for commutation rule monoids (see Cartier and Foata 1969). Content, Lemay, and Leroux (1980) proposed a general setting for Möbius inversion that includes locally finite partially ordered sets and commutation rule monoids. The second fundamental transformation was derived by Foata (1968) and further used in Foata (1977) and Foata and Schützenberger (1978). See MacMahon (1913, 1915, 1916) for the first studies of the major index. Several multivariate distributions on $\mathfrak{S}_n$ involving the major index and inversion number have been calculated, particularly by Stanley (1976), Gessel (1977), Garsia and Gessel (1979), Rawlings (1981). The extension of the equivalence principle is due to Sparre-Andersen (1962). Other combinatorial constructions have been found that basically involve rearrangements of sequences. See for example Dumont and Viennot (1980), Dumont (1981) and Strehl (1981).

Problems

Section 10.2

10.2.1. For $0 \leqslant k \leqslant n$ let $A_{n,k}$ denote the number of permutations in $\mathfrak{S}_n$ having k descents. Take a permutation $w = a_1 a_2 \cdots a_{n-1}$ ($n \geqslant 2$) and insert n before w, after w or between two letters. The number of descents remains alike or increases by one. This provides the recurrence relation for the *Eulerian numbers* $A_{n,k}$, that reads

$$A_{1,0} = 1, \qquad A_{1,k} = 0 \quad \text{for} \quad k \neq 1$$

and for $n \geqslant 2$ and $0 \leqslant k \leqslant n - 1$

$$A_{n,k} = (k+1)A_{n-1,k} + (n-k)A_{n-1,k-1}.$$

(See Foata and Schützenberger 1970).

10.2.2. For each permutation $w = a_1 a_2 \cdots a_n$ the *number of rises* of w, denoted by $R(w)$, is defined to be the number of integers j with $0 \leqslant j \leqslant n - 1$ and $a_j < a_{j+1}$ (by convention $a_0 = 0$), while the *number of 0-exceedances* of w, denoted by $E_0(w)$, is the number of integers j with $1 \leqslant j \leqslant n$ and $a_j \geqslant j$. Note that $E_0 \neq E + 1$.

Consider the following sequence

$$w = a_1 a_2 \cdots a_n,$$
$$w_1 = a_2 a_3 \cdots a_n a_1$$
$$w_2 = \hat{w}_1 \quad (\text{``first fundamental transformation''})$$
$$w_3 = \tilde{w}_2 \quad (\text{reverse image}).$$

The mappings $w \mapsto w_3$ and $w_2 \mapsto w_3$ are bijections of $\mathfrak{S}_n$ onto itself with the property that

$$E_0(w) = R(w_3) = (1 + D)(w_2).$$

(See Foata and Schützenberger 1970.)

10.2.3. For each positive integer n let

$$A_n(t) = \sum A_{n,k} t^k \quad (0 \leqslant k \leqslant n - 1)$$

be the nth *Eulerian polynomial*. From Problem 10.2.1 it follows that

$$A_n(t) = \sum t^{D(w)} \quad (w \in \mathfrak{S}_n),$$

and from Problem 10.2.2

$$t A_n(t) = \sum t^{R(w)} \quad (w \in \mathfrak{S}_n).$$

By classifying the permutations according to the position of the letter n we have

$$A_n(t) = A_{n-1}(t) + t \sum \binom{n-1}{m} A_m(t) A_{n-1-m}(t)$$

$$(0 \leqslant m \leqslant n-2; \quad n \geqslant 1)$$

and

$$t A_n(t) = \sum \binom{n-1}{m} t A_m(t) \cdot t A_{n-1-m}(t)$$

$$(0 \leqslant m \leqslant n-1; \quad n \geqslant 1)$$

The former identity is equivalent to

$$1 + \sum A_n(t) u^n / n! = \exp\left(u + \sum t A_{m-1}(t) u^m / m!\right)$$

$$(n \geqslant 1; \quad m > 2),$$

whereas the latter one is equivalent to

$$1 + \sum t A_n(t) u^n / n! = \exp \sum t A_{n-1}(t) u^n / n! \quad (n \geqslant 1).$$

The last two identities form a system of two equations with two unknowns. Solving this system yields

$$1 + \sum A_n(t) u^n / n! = (1 - t) / (-t + \exp(ut - u))$$

$$1 + \sum t A_n(t) u^n / n! = (1 - t) / (1 - t \exp(u - ut)) \quad (n \geqslant 1).$$

Section 10.4

10.4.1. Let Π be the bijection of the circuit monoid $C(A)$ onto A^* and Γ be the inverse bijection, as they were defined in Section 10.4. For each pair of words w, w' in A^* the formula $w\tau w' = \Pi(\Gamma(w)\Gamma(w'))$ defines a new product in A^*, called the *intercalation product*. The ordered pair $C'(A) = (A^*, \tau)$ is called the *intercalation monoid*. It is isomorphic to $C(A)$. Let w, w' be two words and denote by $a_1, a_2, \ldots, a_n$ the increasing sequence of the letters occurring in either w or w'. Let $m_i = |w|_{a_i}$ (resp. $m_i' = |w'|_{a_i}$) be the number of occurrences of a_i in w (resp. in w') and $(w_1, w_2, \ldots, w_n)$ (resp. $(w_1', w_2', \ldots, w_n')$) be the factorization of w defined by $|w_i| = m_i$ (resp. of w' defined by $|w_i'| = m_i'$). Then

$$w\tau w' = w_1 w_1' w_2 w_2' \cdots w_n w_n'.$$

For instance, with $w = 311454$ and $w' = 52243$ we have $w\tau w' = 31521245443$. (See Cartier and Foata 1969.)

10.4.2. A *cycle* is defined to be a nonempty circuit

$$c = \begin{bmatrix} \delta w \\ w \end{bmatrix} = \begin{bmatrix} a_2 a_3 \cdots a_m & a_1 \\ a_1 a_2 \cdots a_{m-1} & a_m \end{bmatrix} \quad \text{with} \quad w = a_1 a_2 \cdots a_m$$

standard. Two cycles $c = \begin{bmatrix} \delta w \\ w \end{bmatrix}$ and $c' = \begin{bmatrix} \delta w' \\ w' \end{bmatrix}$ are said to be *disjoint* if w and w' have no letter in common. The circuit monoid $C(A)$ is generated by the set of all cycles submitted to the following commutation rule that $cc' = c'c$ whenever c and c' are disjoint. (See Cartier and Foata 1969.)

10.4.3. Let n be a positive integer and A be the finite alphabet $\{1, 2, \ldots, n\}$. Construct the circuit monoid $C(A)$. If a circuit c is a product of exactly $p(c)$ disjoint cycles, let

$$\mu(c) = (-1)^{p(c)}.$$

In the other cases, let $\mu(c) = 0$. The *characteristic series* of $C(A)$ is given by

$$\sum c = \left(\sum \mu(c) c \right)^{-1} \quad (c \in C(A)).$$

(See Cartier and Foata 1969.)

10.4.4. Let $B = (b(i, j))$ $(1 \leqslant i, j \leqslant n)$ be an $n \times n$ matrix. We assume that the n^2 entries $b(i, j)$ are indeterminates subject to the following commutation rule that $b(i, j)$ and $b(i', j')$ commute whenever i and

i' are distinct. Let $\mathbb{Z}[[B]]$ denote the $\mathbb{Z}$-algebra of formal power series in the variables $b(i, j)$s (still subject to the foregoing commutation rule). The polynomial $\det(I - B)$ (with I the identity matrix of order n) belongs to $\mathbb{Z}[[B]]$. For each nonempty circuit

$$c = \begin{bmatrix} a_1' & a_2' & \cdots & a_m' \\ a_1 & a_2 & \cdots & a_m \end{bmatrix}$$

let

$$\beta(c) = b(a_1', a_1) b(a_2', a_1) \cdots b(a_m', a_m)$$

and $\beta(c) = 1$ if c is the empty circuit. The following identity holds:

$$\det(I - B) = \sum \mu(c)\beta(c).$$

By extending β to a homomorphism of the large algebra of $C(A)$ into $\mathbb{Z}[[B]]$ we deduce from Problem 10.4.3 that

$$\left(\det(I - B)\right)^{-1} = \sum \beta(c) \qquad (c \in C(A)).$$

(See Cartier and Foata 1969.)

10.4.5. Let $X_1, X_2, \ldots, X_n$ be n commuting variables, and let $B' = (b_{ij}')$ $(1 \leqslant i, j \leqslant n)$ be a matrix with real entries. Let $\alpha(b(i, j)) = b_{ij}' X_j$ and extend the definition of α to all of $\mathbb{Z}[[B]]$ by linearity. The image under α of the latter identity is

$$\begin{vmatrix} 1 - b_{11}' X_1 & \cdots & - b_{1n}' X_n \\ \cdots & \cdots & \cdots \\ - b_{n1}' X_1 & \cdots & 1 - b_{nn}' X_n \end{vmatrix}^{-1} =$$

$$\sum a(m_1, m_2, \ldots, m_n) X_1^{m_1} X_2^{m_2} \cdots X_n^{m_n}$$

$$(m_1 \geqslant 0, m_2 \geqslant 0, \ldots, m_n \geqslant 0),$$

where $a(m_1, m_2, \ldots, m_n)$ is the coefficient of the monomial $X_1^{m_1} X_2^{m_2} \cdots X_n^{m_n}$ in the expansion of

$$\left(\sum_j b_{1j}' X_j\right)^{m_1} \left(\sum_j b_{2j}' X_j\right)^{m_2} \cdots \left(\sum_j b_{nj}' X_j\right)^{m_n}.$$

This identity constitutes the essence of the MacMahon Master Theorem. (See Cartier and Foata 1969.)

Section 10.6

10.6.1. The algorithm for the inverse Φ^{-1} of the second fundamental transformation can be described as follows:

Let $w = a_1 a_2 \cdots a_m$ be a word.
1. Let $i = m$, $v' = a_1 a_2 \cdots a_m$;
2. Let b_i be the last letter of v'; if $i = 1$, let $\Phi^{-1}(w) = b_1 b_2 \cdots b_m$, else let v_{i-1} be defined by $v' = v_{i-1} b_i$;
3. If the first letter of v_{i-1} is greater than (resp. smaller than or equal to) b_i, split v_{i-1} before each letter greater than (resp. smaller than or equal to) b_i;
4. In each compartment of v_{i-1} determined by the splits move the first letter to the end; let v' be the word obtained after making those moves; replace i by $i - 1$ and go to 2.

(See Foata 1968; Foata and Schützenberger 1978.)

10.6.2. Let q be a real or complex variable and for each positive integer m let $[m] = 1 + q + q^2 + \cdots + q^{m-1}$ and $[m] = 1$ if $m = 0$. Also let $[m]! = [m][m-1] \cdots [2][1]$. Let X be the rearrangement class of the word $1^{m_1} 2^{m_2} \cdots n^{m_n}$. Then

$$\frac{[m_1 + m_2 + \cdots + m_n]!}{[m_1]![m_2]! \cdots [m_n]!} = \sum q^{\mathrm{INV}\, w} = \sum q^{\mathrm{MAJ}\, w} \quad (w \in X).$$

(See Andrews 1976, Chapter 3).

10.6.3. For each $w = a_1 a_2 \cdots a_n$ in the permutation group $\mathfrak{S}_n$ let IDOWN w be the down set of the inverse w^{-1} (in the group $\mathfrak{S}_n$). Clearly, the integer i belongs to IDOWN w if and only if in the word $a_1 a_2 \cdots a_n$ the letter $i + 1$ occurs to the left of the letter i. Let IMAJ w be the sum of the elements in IDOWN w. The second fundamental transformation, *restricted to the permutation group $\mathfrak{S}_n$*, preserves IDOWN; that is:

$$\mathrm{IDOWN}\, \Phi(w) = \mathrm{IDOWN}\, w.$$

Denote by $\mathbf{i}(w)$ the inverse w^{-1} of the permutation w and consider the transformation

$$\Psi = \mathbf{i}\Phi\mathbf{i}\Phi^{-1}\mathbf{i}.$$

Then Ψ is a bijection of $\mathfrak{S}_n$ onto itself with the property that

$$\mathrm{MAJ}\, \Psi(w) = \mathrm{INV}\, w \quad \text{and} \quad \mathrm{INV}\, \Psi(w) = \mathrm{MAJ}\, w.$$

In particular, the six ordered pairs (MAJ, INV), (IMAJ, INV), (IMAJ, MAJ), (MAJ, IMAJ), (INV, IMAJ), and (INV, MAJ) have the same bivariate distribution on $\mathfrak{S}_n$. (See Foata and Schützenberger 1978.)

CHAPTER 11

Words and Trees

11.0. Introduction

The aim of this chapter is to give a detailed presentation of the relation between plane trees and special families of words: parenthesis systems and other families. The relation between trees and parenthesis notation is classical and has been known perhaps since Catalan 1838.

Because trees play a central role in the field of combinatorial algorithms (Knuth 1968), their coding by parenthesis notation has been investigated so very often that it is quite impossible to give a complete list of all the papers dealing with the topic. These subjects are also considered in enumeration theory and are known to combinatorialists (Comtet 1970) as being counted by Catalan numbers. Note that a generalization of the type of parenthesis system often called Dyck language is a central concept in formal language theory. These remarks give a good account of the main role played by trees and their coding in combinatorics on words.

Presented here are three ways to represent trees by words. The first one consists in constructing a set of words (one for each node) associated to a plane tree. The second is the classical parenthesis coding, and the third concerns Lukaciewicz language (known also as Polish notation).

The combinatorial properties of Lukaciewicz language were investigated by Raney (1960) in order to give a purely combinatorial proof of the Lagrange inversion formula (see also Schützenberger 1971). This proof is presented in Section 11.4 of the present chapter as an application of our combinatorial constructions.

Among the many ways of defining trees, the most usual is to say that a tree is a connected graph with no cycles. The preferred definition here uses a characteristic property of mapping that assigns to any node the set of its "sons." This definition allows easy generalization for the introduction of plane trees. These definitions are given in the first section, which ends with Dewey notation.

In the second section the parenthesis coding of a tree is introduced as a consequence of the canonical decomposition of a tree into two subtrees. In the third properties of Lukaciewicz languages are investigated and another coding for trees is constructed. In the fourth Raney's proof of Lagrange inversion formula is presented.

11.1.   Trees and Plane Trees

Let S be a finite set, its elements to be called *nodes*; let r be a distinguished element of S. A *tree* with *root* r is a mapping α from S into the set $\mathcal{P}(S)$ of all subsets of S satisfying the following condition:

(T): For any s in S there exists a unique sequence $(s_1, s_2, \ldots, s_p)$ $(p \geqslant 1)$ such that $s_1 = r$, $s_p = s$, and $s_{i+1} \in \alpha(s_i)$ for $i = 1, p - 1$.

Such a sequence will be called a *path* from r to s. The condition (T) is very strong, particularly the uniqueness part. It implies for instance that r does not belong to any $\alpha(s)$ and that for any s, $s \neq r$ there exists a unique t with $s \in \alpha(t)$. Also for any sequence $s_1, s_2, \ldots, s_p$ such that $s_{i+1} \in \alpha(s_i)$ for $i = 1, \ldots, p - 1$ one has $s_1 \notin \alpha(s_p)$. This last remark allows us to prove the following proposition, the complete proof of which is left to the reader.

A *leaf* in a tree α is a node s for which $\alpha(s) = \varnothing$; given a subset T of S the restriction α_T of α to T is defined by $\alpha_T(t) = \alpha(t) \cap T$, for any t in T.

PROPOSITION 11.1.1. *Let α be a tree on the set S of nodes with root r; then there exists at least one leaf s_f in S. Moreover if S contains more than one element, the restriction of α to $S \setminus \{s_f\}$ is still a tree.*

From that result many properties on trees may be proved using induction on the cardinality n of the set S of nodes; let us as an illustration prove the following one:

PROPOSITION 11.1.2. *If α is a tree on a set S of cardinality n, then:*

$$\sum_{s \in S} \operatorname{Card} \alpha(s) = n - 1.$$

Proof. The result is trivially obtained when $n = 1$ since in that case $S = \{r\}$ and $\alpha(r) = \varnothing$. If (S, α) is a tree with $n + 1$ nodes, then by property 11.1.1, there exists s_f with $\alpha(s_f) = \varnothing$, and α', the restriction of α to $S \setminus \{s_f\}$, is a tree. The induction hypothesis implies

$$\sum_{s} \operatorname{Card} \alpha'(s) = n - 1.$$

But $\alpha'(s) = \alpha(s)$ except for the unique t verifying $s_f \in \alpha(t)$, for which $\alpha(t) = \alpha'(t) \cup \{s_f\}$. This ends the proof.   ∎

A sequence $(s_1, s_2, \ldots, s_p)$ of elements of S is said to be *proper* if $s_i = s_j$ implies $i = j$. A *plane tree* on the set of nodes S with root r is a mapping ϕ into the set of proper sequences of elements of S such that the mapping $\tilde{\phi}$ from S into $\mathcal{P}(S)$ induced by ϕ ($t \in \tilde{\phi}(s)$ if t occurs in $\phi(s)$)) verifies condition (T).

For any tree α, a *representation* of α is a plane tree ϕ such that $\tilde{\phi} = \alpha$; if n_s denotes for each s the cardinality of the set $\alpha(s)$ then the number of representations of the tree α is equal to the product

$$\prod_{s \in S} n_s!$$

Example 11.1.3. The tree α defined on $S = \{1, 2, 3, 4, 5, 6, 7\}$ by $\alpha(1) = \{2\}$; $\alpha(2) = \{3, 4, 5\}$; $\alpha(3) = \{6, 7\}$, and $\alpha(i) = \emptyset$ for $i \geq 4$ has 12 representations, one of which is $\phi(1) = 2$; $\phi(2) = (4, 3, 5)$; $\phi(3) = (7, 6)$; $\phi(i) = \emptyset$ for $i > 4$. See Figure 11.1.

From now on, the discussion will deal only with plane trees, so henceforth in this chapter the word *tree* will be an abbreviation for *plane tree*.

If A is an alphabet on which is defined a total order $<$, recall that the lexicographic order (denoted also by $<$) on the free monoid A^* extends $<$ in the following way:

If $(u, v) \in A^*$, then $u < v$ if either $v \in uA^+$ or $u = ras, v = rbt$, where a and b are elements of A such that $a < b$. For each u in A let us denote by $\inf(u)$ the set of words v such that $u \in vA^*$ or $u = u'a$, $v = u'b$, $a, b \in A$ and $b < a$. Remark that if v belongs to $\inf(u)$, $v < u$ and $|v| \leq |u|$, but the converse is not true: if we take $A = \{a, b\}b < a$, then $abb < aab$ but $abb \notin \inf(aab)$. We say that a subset C of A^* is *closed* if for any u in C $\inf(u) \subset C$. For any subset C of A^* let $\phi_C(u)$ (for $u \in C$) be the sequence of elements $(u_1, u_2, \ldots, u_p)$ of $C \cap uA$ in increasing order $(u_1 < u_2 \cdots < u_p)$.

PROPOSITION 11.1.4. *For any closed subset C of A^*, the mapping ϕ is a tree (that is, plane tree) with root 1. Conversely for any tree ϕ on the set S there*

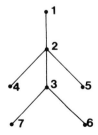

Figure 11.1. A representation of a tree.

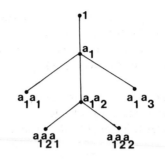

Figure 11.2. The subset associated to the tree ϕ of Example 11.1.3.

*exists an alphabet A, a closed subset C of A^*, and a bijection β from S onto C
such that*

$$\phi(s) = (s_1, s_2, \dots, s_p) \Leftrightarrow \tilde{\phi}_C(\beta(s)) = (\beta(s_1), \beta(s_2), \dots, \beta(s_p))$$

(Dewey notation).

Proof. Let f be an element of C, $f = a_1 a_2 \cdots a_p$; then $1 < a_1 < a_1 a_2 < \cdots < a_1 a_2 \cdots a_{p-1} < f$. Denote by g_i the word $a_1 a_2 \cdots a_i$; then clearly g_{i+1} belongs to $\tilde{\phi}_C(g_i)$, and $(1, g_1, g_2, \dots, g_{p-1}, f)$ is a path from 1 to f. Now if $(1, h_1, h_2, \dots, h_q = f)$ is a sequence with $h_{i+1} \in \tilde{\phi}_C(h_i)$, then $|h_i| = i$, $q = p$, and by definition of $\tilde{\phi}_C$, $h_p = h_{p-1} a_p$, thus $h_{p-1} = g_{p-1}$, and by induction, we obtain $g_i = h_i$ for every i. We have thus proved that $\tilde{\phi}_C$ is a tree.

Conversely, let ϕ be a tree on a set S with root r; denote by $u = (s_1, s_2, \dots, s_q)$ the longest sequence among all the $\phi(s)$, and let A be the alphabet $\{a_1, a_2, \dots, a_q\}$ ordered by $a_1 < a_2 \cdots < a_q$. For any s in S, condition (T) ensures the existence of a path $(t_1 = r, t_2, \dots, t_j, \dots, t_p = s)$ from r to s; this allows the definition of the mapping β from S to A^* by

$$\beta(s) = a_{i_1} a_{i_2} \cdots a_{i_p},$$

where i_j is such that t_{j+1} is the i_jth element in the sequence $\phi(t_j)$.

It is not difficult to verify that $\beta(S)$ is a closed subset of A^* and that β satisfies the conditions of Proposition 11.1.4.   ∎

Example 11.2.5. The subset $C = \beta(S)$ associated to the tree ϕ of Example 11.1.3 is $A = \{a_1, a_2, a_3\}$; $U = \{1, a_1, a_1 a_1, a_1 a_2, a_1 a_3, a_1 a_2 a_1, a_1 a_2 a_2\}$. See Figure 11.2.

11.2.   Trees and Parenthesis Systems

The "coding" of a tree by a parenthesis system follows from the fact that a tree can be decomposed into subtrees in the same way as parenthesis systems can be decomposed into subsystems.

Let us first examine the decomposition of trees. Let ϕ be a tree on the set S with root r. If S is not reduced to $\{r\}$, then $\phi(r)$ is not empty. Denoting then by $(r_1, r_2, \ldots, r_q)$ the sequence $\phi(r)$, one has $q \geq 1$. Let us consider the subsets $S_0, S_1, \ldots, S_q$ of S defined by $S_0 = \{r\}$ and S_i (for $1 \leq i \leq q$), which is the set of all the nodes s such that the path $(s_0 = r, s_1, \ldots, s_p = s)$ from r to s is such that $s_i = r_i$. Clearly, these subsets define a partition of S, and if s is in the subset S_i, then this also holds for any t in $\tilde{\phi}(s)$.

Moreover, the restriction ϕ_i of ϕ to the subset S_i is a tree with root r_i; we will denote by $F_i(\phi)$ the tree ϕ_i. If we put $T = S_0 \cup S_2 \cdots \cup S_q$ then the restriction of ϕ to T is also a tree, which will be denoted by $G(\phi)$.

The following two propositions state that the $F_i(\phi)$ and $(F_1(\phi), G(\phi))$ are sufficient to reconstruct ϕ. Their proofs are left as exercises.

PROPOSITION 11.2.1. *The pair* $(F_1(\phi), G(\phi))$ *of trees defined on two disjoint subsets uniquely determines the tree* ϕ *defined on the union of these subsets.*

PROPOSITION 11.2.2. *The sequence* $(F_1(\phi), F_2(\phi), \ldots, F_q(\phi))$ *of trees uniquely determines the tree* ϕ.

Consider as isomorphic two trees differing only by a renaming of their nodes, and let a_n be the number of nonisomorphic plane trees, with n vertices. Then as a consequence of Proposition 11.2.1 we have

$$a_n = \sum_{p=1}^{n-1} a_p a_{n-p}.$$

Denoting by $a(x)$ the generating power series

$$\sum_{n \geq 0} a_n x^n$$

we obtain

$$a(x) = x + (a(x))^2,$$

and expanding $1 - (1 - 4x)^{1/2}$, we have the following:

COROLLARY 11.2.3. *The number of plane trees with n vertices is the Catalan number*:

$$\frac{(2n-2)!}{n!(n-1)!}$$

Let us now consider the alphabet $\{a, \bar{a}\}$ and let δ be the morphism of $\{a, \bar{a}\}^*$ into the additive group $\mathbb{Z}$ of rational integers defined by $\delta(\bar{a}) = -1$ and $\delta(a) = 1$. A *parenthesis system* is a word f of $\{a, \bar{a}\}^*$ such that $\delta(f) = 0$

and $\delta(f') \geqslant 0$ for any left factor f' of f. Let P denote the set of parenthesis systems.

PROPOSITION 11.2.4. *Any word f of P different from 1 has a unique decomposition $f = a\, f_1\, \bar{a}\, f_2$ with $f_1, f_2 \in P$. Conversely, if $f_1, f_2 \in P$ then $a\, f_1\, \bar{a}\, f_2$ is also an element of P.*

Proof.

Let f be an element of P and let g be the shortest nonempty left factor of f whose image by δ is zero. Clearly, g begins with a and ends with $\bar{a}$; if not, f would possess a left factor h such that $\delta(h) = -1$. Denote $g = a\, f_1\, \bar{a}$ and $f = a\, f_1\, \bar{a}\, f_2$; it is then easy to verify that $\delta(f_1) = \delta(f_2) = 0$ and that any left factor of f_1 or f_2 has a nonnegative image by δ (use the minimal length of g and the fact that $f \in P$). Now if f has two decompositions, $f = a\, f_1\, \bar{a}\, f_2 = a\, g_1\, \bar{a}\, g_2$; $f_1 \in P$ and $\delta(g_1\bar{a}) = -1$ imply $|f_1| \geqslant |g_1|$. Symmetrically $|f_1| \leqslant |g_1|$ and thus $f_1 = g_1$, $f_2 = g_2$, which proves the first part of the proposition. The converse is easy to establish and is left as exercise.   ∎

Comparing Propositions 11.2.1 and 11.2.4 one can define a mapping Π from the set of trees on S onto the set of parenthesis systems recursively on the size of S_i.

If $S = \{r\}$ then the only tree ϕ_0 on S verifies $\phi_0(r) = \varnothing$ and $\Pi(\phi_0)$ is the empty word 1.

Let Π be defined on the set of trees with less than n nodes; if ϕ is a tree with n nodes let

$$\Pi(\phi) = a\Pi(F_1(\phi))\bar{a}\Pi(G(\phi))$$

THEOREM 11.2.5. *To any tree ϕ with n nodes Π associates a parenthesis system of length $2n - 2$. Moreover Π is surjective, and two trees with the same image by Π differ only by a renaming of the set of their nodes.*

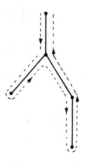

a a ā a a ā ā ā

Figure 11.3. The parenthesis system associated with a tree.

The proof by induction is a direct consequence of Propositions 11.2.1 and 11.2.3.

Example 11.2.6. The parenthesis system $f = a\,a\,\bar{a}\,a\,a\,\bar{a}\,\bar{a}\,\bar{a}$ is associated with the tree shown in Figure 11.3.

11.3. Lukaciewicz Language

In this section and the following, A will denote the infinite alphabet $A = \{a_0, a_1, \ldots, a_n, \ldots\}$ and δ the morphism of A^* into the additive group $\mathbb{Z}$ of rational integers defined by

$$\delta(a_n) = n - 1.$$

The Lukaciewicz language L is the set of words f of A^* such that $\delta(f) = -1$ and $\delta(f') \geq 0$ for any left factor f' of f. Let us now investigate a few combinatorial properties of this language, properties that will be used in Section 11.4 for the Lagrange formula.

LEMMA 11.3.1. *Let f be a word with $\delta(f) = -p$, $p > 0$; then, for any q, $0 \leq q \leq p$ there exists a left factor f' of f such that $\delta(f') = -q$.*

Proof. (by induction on the length of f). If f is of length 1, then $f = a_0$, $p = -1$, and 1, a_0 are two left factors of f with $\delta(1) = 0, \delta(a_0) = -1$. Let f be of length $n > 1$; then $f = ga_i$; if $i > 0$ the existence of f' is obtained by the inductive hypothesis applied to g as in $\delta(g) = \delta(f) - (i-1) \leq -p$. If $f = ga_0$, then either $q = p$ and then $f' = f$ satisfies $\delta(f') = -q$, or $q \leq p - 1$ and the inductive hypothesis applied to $g, (\delta(g) = -p + 1)$ gives the result. ∎

PROPOSITION 11.3.2. *Any word of L has a unique decomposition as a product of words of L.*

Proof. This is a direct consequence from the fact that L is prefix. More precisely, assume that it is not true, and let f be the shortest word having two distinct decompositions as a product of words of L.

$$f = g_1 \cdots g_p = h_1 \cdots h_q.$$

Then by the minimality of f one has $g_1 \neq h_1$, and one of the two has the other as a left factor, in contradiction with

$$\delta(g_1) = \delta(h_1) = -1, \quad g_1 \in L, \quad h_1 \in L.$$

PROPOSITION 11.3.3. *A word f is in L^p if and only if $\delta(f) = -p$ and $\delta(f') > -p$ for any left factor $f'(f' \neq f)$ of f.*

Proof. If $f = g_1 \cdots g_p$ is in $L^p(g_i \in L)$ then clearly

$$\delta(f) = \sum_{i=1}^{p} \delta(g_i) = -p$$

and for any left factor f' of f one has

$$f' = g_1 \cdots g_{i-1} g_i'$$

where $g_i = g_i' g_i''$ and $1 \leqslant i \leqslant p$. Thus $\delta(f') = -(i-1) + \delta(g_i')$ and g_i' as a left factor of a word of L satisfies $\delta(g_i') \geqslant 0$; thus $\delta(f') \geqslant 1 - i \geqslant -p + 1$.

Conversely, let f be such that $\delta(f) = -p$ and $\delta(f') > -p$ for any left factor f' of f, then by the Lemma 11.3.1, f has a left factor of which the image by δ is $-q$, and this for any $q < p$. Let f_q, $(q \leqslant p)$ be the shortest left factor with image by δ equal to $-q$. Then, also by Lemma 11.3.1, f_i is a left factor of f_{i+1} for any i. Let us write $f_{i+1} = f_i g_{i+1}$ and $f_1 = g_1$; we obtain $f = g_1 g_2 \cdots g_p$. Clearly $\delta(g_i) = \delta(f_{i+1}) - \delta(f_i) = -1$ and any left factor g_i' of g_i verifies $\delta(g_i') \geqslant 0$, because, if not, $f_i g_i'$ by Lemma 11.3.1 would have a left factor f_i' such that $\delta(f_i') = -(i+1)$ in contradiction with the minimality of f_{i+1}. ∎

PROPOSITION 11.3.4. *Any word f of L has a unique decomposition $f = a_k f_1 \cdots f_k$ with $f_i \in L$.*

Let f be in L, then $f = a_k g$ with a_k in A. Clearly g verifies $\delta(g) = \delta(f) - \delta(a_k) = -1 - (k-1) = -k$; and for any left factor g' of g, $a_k g'$ is a left factor of f; then $\delta(a_k g') \geqslant 0$ and $\delta(g') > -k$. Thus by Proposition 11.3.3, g belongs to L^k and by 11.3.2, g has a unique decomposition $g = g_1 \cdots g_k$ yielding the decomposition of f. ∎

From Propositions 11.2.2 and 11.3.4 we can construct a mapping Λ from the set of trees onto L in the same way as Π was constructed in Section 11.2:

$$\Lambda(\phi_0) = a_0.$$

If Λ is defined on the set of trees with fewer than n nodes, then:

$$\Lambda(\phi) = a_q \Lambda(F_1(\phi)) \Lambda(F_2(\phi)) \cdots \Lambda(F_q(\phi)).$$

We have also:

THEOREM 11.3.5. *If ϕ has n nodes $\Lambda(\phi)$ has length n. Moreover Λ is surjective, and two trees with the same image by Λ differ only by a renaming of the set of their nodes.*

A factorization of a word f is a pair (f_1, f_2) of words such that $f_1 \neq 1$ and $f = f_1 f_2$. A word f has then exactly $|f|$ factorizations.

THEOREM 11.3.6. *Any word f, with $\delta(f) = -p$ ($p > 0$) has exactly p factorizations (f_1, f_2) such that $f_2 f_1 \in L^p$.*

Let us first prove the result for a word $f = g_1 g_2 \cdots g_p$ of L^p; clearly, $(g_1, g_2 \cdots g_p), (g_1 g_2, g_3 \cdots g_p) \cdots (g_1 g_2 \cdots g_{p-1}, g_p)$, and $(g_1 \cdots g_p, 1)$ are p such factorizations. Let us show that these are the only ones. If (f_1, f_2) is such that $f_2 f_1 \in L^p$ and $f = f_1 f_2$, then $f_1 = g_1 g_2 \cdots g_{i-1} g_i'$, $f_2 = g_i'' g_{i+1} \cdots g_p$, and $g_i = g_i' g_i''$; then $f_2 f_1 = g_i'' g_{i+1} \cdots g_p g_1 \cdots g_{i-1} g_i'$. Because g_i' is a left factor of a word of L, $\delta(g_i') \geq 0$ and $\delta(g_i'' g_{i+1} g_p \cdots g_{i-1}) \leq -p$. This contradicts Proposition 11.3.3 unless $g_i' = 1$, and this gives $f_1 = g_1 \cdots g_{i-1}$.

Now let f be any word with $\delta(f) = -p$ and, among the left factors of f with minimal image by δ, let g be the shortest; then $f = gh$ and $\delta(g) \leq -p$. Let us consider the word hg; clearly $\delta(hg) = -p$ and any left factor of hg is either a left factor h' of h or of the form hg', where g' is a left factor of g. But $\delta(h') \geq 0$ by the minimality of $\delta(g)$, and $\delta(hg') \geq \delta(g')$ is greater than $-p$ for the same reason.

Thus hg verifies the sufficient conditions of Proposition 11.3.3 and is then an element of L^p. Remarking that the factorizations of hg yield factorization of gh and using the first part of our proof, we obtain the theorem. ∎

Observe that Theorem 11.3.6 is a direct consequence of Theorem 5.4.1 considering the bisection of A^* defined by $X_1 = L$ and $X_2 = L' \cup (A \setminus \{a_0\})$, where $L' = \{f \mid fa_0 \in L\}$.

11.4. Lagrange Inversion Formula

This section is devoted to a combinatorial proof of the Lagrange formula. It begins with statements of a few properties of a morphism U from A^* (where A is the alphabet considered in Section 11.3) into a field K; this morphism is defined for any formal power series u of $K((t))$.

Let K be a (commutative) field, $K[[t]]$ denotes the ring of formal power series in the indeterminate t. A series u will be denoted by $\sum_{i \geq 0} u_i t^i$ the coefficient u_i of t^i in u, will also be written $\langle u, t^i \rangle$.

To any series u is associated the morphism U from A^* into K considered as a monoid for the multiplication by

$$U(a_i) = u_i.$$

This morphism can be extended to finite subsets of A^* by

$$U(B) = \sum_{f \in B} U(f).$$

Clearly, the image by U of the union of disjoint subsets B_1 and B_2 of A^* is the sum $U(B_1) + U(B_2)$, and if any word in $B_1 B_2$ has a unique decomposition as a product of an element of B_1 by an element of B_2, then $U(B_1 B_2) = U(B_1)U(B_2)$.

Let $\hat{U}$ be the mapping from the set of all subsets of A^* onto $K[[t]]$ defined by

$$\langle \hat{U}(B), t^n \rangle = U(B \cap A^n).$$

Then clearly the foregoing properties hold also for $\hat{U}$. Letting $\delta^{-1}(p)$ denote the set of words f of A^* such that $\delta(f) = p$, we have the following:

PROPOSITION 11.4.1. *Let u^n denote the nth power of u. Then*

$$\langle u^n, t^q \rangle = U(A^n \cap \delta^{-1}(q - n)).$$

Proof. Clearly

$$\langle u^n, t^q \rangle = \sum_{i_1 + i_2 + . + i_n = q} \left(u_{i_1} u_{i_2} \cdots u_{i_n} \right)$$

which is also

$$\sum_{i_1 + i_2 + .. + i_n = q} U\left(a_{i_1} a_{i_2} \cdots a_{i_n} \right).$$

But $i_1 + i_2 + \cdots + i_n$ is equal to $\delta(a_{i_1} a_{i_2} \cdots a_{i_n}) + n$ thus $\langle u^n, t^q \rangle = \Sigma U(f)$, the sum being extended to all words f of length n with $\delta(f) = q - n$, and this gives the result. ∎

Let us now consider the equation

$$\xi = tu(\xi), \tag{11.4.1}$$

which can be written also as

$$\xi = u_0 t + u_1 t\xi + u_2 t\xi^2 + \cdots + u_n t\xi^n + \cdots$$

Remark that $\xi \to tu(\xi)$ is a "contraction" (11.4.1) in the ultrametric space of formal power series $K[[t]]$. This implies that (11.4.1) has a unique solution

$x(t)$ in $K[[t]]$. The first few terms of $x(t)$ can be computed easily, giving:

$$x(t) = u_0 t + u_1 u_0 t^2 + (u_2 u_0 u_0 + u_1 u_1 u_0)t^3$$
$$+ (u_3 u_0 u_0 u_0 + u_2 u_1 u_0 u_0 + u_2 u_0 u_1 u_0 + u_1 u_2 u_0 u_0 + u_1 u_1 u_1 u_0)t^4 + \cdots$$

Comparing these few terms with the words of length 1 to 4 of L suggests the following proposition: ∎

PROPOSITION 11.4.2. *The unique solution of Eq. (11.4.1) is $\hat{U}(L)$.*

Proof. By Proposition 11.3.4 one has L as the disjoint union of the subsets $a_k L^k$; then the remarks given about $\hat{U}$ imply

$$\hat{U}(L) = \sum_{k \geqslant 0} \hat{U}(a_k)(\hat{U}(L))^k$$
$$= \sum_{k \geqslant 0} u_k t(\hat{U}(L))^k.$$

Clearly $\hat{U}(L)$ verifies Eq. (11.4.1), and is thus its unique solution.   ∎

PROPOSITION 11.4.3. *The following equality holds for $p, n > 0$*

$$nU(L^p \cap A^n) = pU(A^n \cap \delta^{-1}(-p)).$$

Proof. Consider the subset H_n of $A^* \times A^*$ consisting of pairs of words (f, g) such that $fg \in L^p$, $|fg| = n$ and $f \neq 1$. U is extended into a map of $A^* \times A^*$ into K by defining $U(f, g)$ to be $U(fg) = U(f)U(g) = U(g)U(f)$ as K is commutative. Because any word h of length n has exactly n factorizations,

$$U(H_n) = nU(L^p \cap A^n).$$

But by Theorem 11.3.6, for any h in $A^n \cap \delta^{-1}(-p)$ there are exactly p elements (f, g) in H_n such that $gf = h$. Hence $U(H_n) = pU(A^n \cap \delta^{-1}(-p))$, which proves the equality.   ∎

THEOREM 11.4.4. *The unique solution x of Eq. (11.4.1) in $K[[t]]$ satisfies* $n\langle x, t^n \rangle = \langle u^n, t^{n-1} \rangle$.

Proof. By Proposition 11.4.2 one has

$$\langle x, t^n \rangle = U(L \cap A^n).$$

Then considering the case $p = 1$ in Proposition 11.4.3 gives

$$n\langle x, t^n \rangle = U(A^n \cap \delta^{-1}(-1)).$$

Further, Proposition 11.4.1 with $q - n = -1$ yields

$$U(A^n \cap \delta^{-1}(-1)) = \langle u^n, t^{n-1} \rangle,$$

which proves the theorem.   ∎

Let K be the field $\mathbb{C}$ of complex numbers; the Lagrange formula is generally presented in the equivalent following form:

COROLLARY 11.4.5. *Let $u(z)$ be a function of the complex variable z analytic in and inside a contour C surrounding a point a. Let t be such that equation $\xi = a + tu(\xi)$ has a unique solution x inside C. Then the expansion of x is given by*

$$x(t) = a + \sum_{n \geq 1} \frac{t^n}{n!} \left[\frac{d^{n-1}}{dz^{n-1}} u^n(z) \right]_{z = a}.$$

Proof. We may assume that $a = 0$. In fact, the general case follows from $a = 0$ changing z in $z + a$.

Considering x as a formal power series in t we obtain from Theorem 11.4.4

$$\langle x, t^n \rangle = \frac{1}{n} \langle u^n, z^{n-1} \rangle.$$

But as u^n is analytic, the Taylor expansion of u^n gives

$$\langle u^n, z^{n-1} \rangle = \frac{1}{(n-1)!} \left[\frac{d^{n-1}}{dz^{n-1}} u^n(z) \right]_{z = 0},$$

and the result holds.

Theorem 11.4.4 can be generalized in the following way:

THEOREM 11.4.6. *Let F be any formal power series of $[[t]]$. The solution $x(t)$ of Eq. (11.4.1) verifies for $n > 0$:*

$$n\langle F(x), t^n \rangle = \langle F'(t) u^n, t^{n-1} \rangle.$$

Proof. Let us denote by f_p the coefficient $\langle F(t), t^p \rangle$, then since $\langle x^p, t^n \rangle = 0$ if $p > n$, one has:

$$\langle F(x), t^n \rangle = \sum_{p=1}^{n} f_p \langle x^p, t^n \rangle.$$

But, by Proposition 11.4.2, $\langle x^p, t^n \rangle = \langle (\hat{U}(L))^p, t^n \rangle$, and because any word in L^p has a unique decomposition as a product of p words of L:

$$\langle x^p, t^n \rangle = U(L^p \cap A^n).$$

Applying Proposition 11.4.3 and taking $q = n - p$ in 11.4.1, we obtain

$$n \langle x^p, t^n \rangle = p \langle u^n, t^{n-p} \rangle$$

and thus

$$n \langle F(x), t^n \rangle = \sum_{p=1}^{n} p f_p \langle u^n, t^{n-p} \rangle$$

$$= \sum_{p=1}^{n} \langle F', t^{p-1} \rangle \langle u^n, t^{n-p} \rangle.$$

The result follows from the definition of the product of two formal power series.   ∎

As for the preceding theorem (11.4.4), there is an analytic function version of the Theorem 11.4.6:

COROLLARY 11.4.7. *Let $F(z)$ be an analytic function of the complex variable; then, under the same assumptions as in Corollary 11.4.5, one has*

$$F(x(t)) = F(a) + \sum_{n \geqslant 1} \frac{t^n}{n!} \left[\frac{d^{n-1}}{dz^{n-1}} F'(z) u^n(z) \right]_{z=a}.$$

Problems

Section 11.1

11.1.1. Prove that if in a tree α with root r, if s is an element of S such that the restriction of α to $S \setminus \{s\}$ is still a tree, then s is a leaf.

11.1.2. A binary tree on S is a tree α in which for any s, Card $\alpha(s) = 0$ or 2. Prove that for a binary tree the number of leaves is (Card $S + 1)/2$.

11.1.3. Let ϕ be a plane binary tree on S and let A be the alphabet $\{a, b\}$. Show that if β is the bijection defined in Proposition 11.1.4, F the sets of leaves of ϕ then $C = \beta(F)$ satisfies

$$c_1, c_2 \in C, \quad u \in A^*, \quad c_1 = c_2 u \Rightarrow c_1 = c_2 \tag{1}$$

$$u \in A^* \Rightarrow \exists c \quad \text{such that} \quad u = c u_1 \quad \text{or} \quad c = u u_1. \tag{2}$$

Section 11.2

11.2.1. Let $A = \{a_0, a_1, \ldots, a_n, \ldots\}$. Show that the morphism θ of A^* onto $\{a, \bar{a}\}$ given by $\theta(a_i) = a^i\bar{a}$ is a bijection from L onto $P\bar{a}$.

Section 11.3

11.3.1. Let ϕ be a binary tree. Prove that $\Lambda(\phi)$ is a word of $P'a_0$ (where P' is obtained from P by setting $a = a_2, \bar{a} = a_0$).

11.3.2. Use Theorems 11.3.5 and 11.3.6 to give the number of plane trees such that $\{s | \mathrm{Card}\, \tilde{\phi}(s) = i\} = d_i$. (See Harary, Prins, and Tutte 1964).

11.3.3. Use Theorem 11.3.6 and Problem 11.2.1 to show that the number of parenthesis systems of length $2n$ is the Catalan number $(2n)!/(n!(n+1)!)$.

11.3.4. Let $A = \{a_0, a_1, \ldots, a_n, \ldots\}$, $B = \{b_0, b_1, \ldots, b_n, \ldots\}$ be two infinite alphabets, and let C be the subset of $A \times B^*$ of pairs (a_k, g); $a_k \in A$ and $|g| = k$. Show that C^* is a free submonoid of $A^* \times B^*$.

For any (f, g) in C^*, $g = b_{i_1} b_{i_2} \cdots b_{i_k}$ define $\Delta(f, g)$ to be

$$-|f| + \sum_{j=1, k} i_j;$$

Let M be the subset of C^* consisting of all pairs (f, g) satisfying $\Delta(f, g) = -1$ and $\Delta(f_1, g_1) \geq 0$ for all $(f_1, g_1) \in C^*$ such that $(f, g) = (f_1 f_2, g_1 g_2)$ $(f_2 \neq 1)$. Construct subsets $C_0, C_1, C_2, \ldots, C_n, \ldots$ of C satisfying

$$M = \sum_{i \geq 0} C_i M^i.$$

Verify that Theorem 11.3.6 holds also for M. Deduce from this theorem that for any (f, g) in C^* satisfying $\Delta(f, g) = -1$, there exist exactly $|g|$ factorizations $(f; g) = (f_1, g_1)(f_2, g_2)$, $g_1 \neq 1$, such that $(f_2 f_1, g_2 g_1) \in M$. (See Chottin 1975).

Section 11.4

11.4.1. Let u and v be formal power series in $K[[t_1]]$ and $K[[t_2]]$, respectively. Let $x(t_1, t_2)$ and $y(t_1, t_2)$ be the solution of the system of equations

$$\xi = t_1 u(\eta) = t_1 \left(u_0 + u_1 \eta + u_2 \eta^2 + \cdots + u_p \eta^p + \cdots \right)$$

$$\eta = t_2 v(\xi) = t_2 \left(v_0 + v_1 \xi + v_2 \xi^2 + \cdots + v_p \xi^p + \cdots \right).$$

Let U and $\hat{U}$ be the morphisms of $A^* \times B^*$ on K and $K[[t_1, t_2]]$, respectively, defined by

$$U(a_i, 1) = u_i \qquad U(1, b_i) = v_i$$
$$\hat{U}(a_i, 1) = u_i t_1 \qquad \hat{U}(1, b_i) = v_i t_2$$

Prove the following identities:

$$\langle u^n, t_1^m \rangle \langle v^m, t_2^q \rangle = U(\Delta^{-1}(q - n) \cap A^n \times B^m) \qquad (1)$$

$$x(t) = \hat{U}(M) \qquad (2)$$

$$n \cup (M \cap A^n \times B^m) = \cup(\Delta^{-1}(-1) \cap A^n \times B^m) \qquad (3)$$

Good formula: $\quad n\langle x, t_1^n t_2^m \rangle = \langle u^n, t_1^m \rangle \langle v^m, t_2^{n-1} \rangle \qquad (4)$

(See Good 1960, Chottin 1975)

11.4.2. A peak of a word f of $\{a, \bar{a}\}^*$ is a factor f of the form $a\bar{a}$ of f. Verify that for any word $f\bar{a}$ satisfying $\delta(f\bar{a}) = -1$, each factorization (f_1, f_2) of f constructed by Lemma 11.3.1 is such that $f_2 f_1$ has as many peaks as f.

Prove that the number of elements of $P\bar{a}$ with p peaks length $2n - 1$ is $1/(n - p)$ the number of elements of $\delta^{-1}(-1) \cap \{a, \bar{a}\}^{2n-3}$ with p peaks.

Deduce that the number of elements of P with p peaks and length $2n$ is

$$\frac{1}{n - p + 1}\binom{n}{p}\binom{n-1}{p-1}.$$

What is the number of plane trees with $n + 1$ nodes, p of which being leaves? (See Narayana 1959; Gouyou Beauchamps 1975).

Bibliography

Bibliography

Adian, S. I., 1979, The Burnside Problem and Identities in Groups. *Ergeb. Math. Grenzgeb.*, *95*, Springer–Verlag, Berlin.

Aho, A. V., J. E. Hopcroft, and J. D. Ullman, 1974, *The Design and Analysis of Computer Algorithms*, Addison-Wesley, Reading, Mass.

Anderson, P. G., 1976, *Amer. Math. Monthly 83*, 359–361.

Andrews, G. E., 1976, *The Theory of Partitions*, Addison-Wesley, Advanced Book Program, Reading, Mass.

Appel, K. I., and F. M. Djorup, 1968, On the equation $z_1^n z_2^n \cdots z_k^n = y^n$ in a free semi-group, *Trans. Am. Math. Soc. 134*, 461–470.

Arson, S., 1937, Proof of the existence of infinite assymetric sequences (Russian), *Mat. Sb. 44*, 769–777.

Assous, R., and M. Pouzet, 1979, Une caractèristation des mots périodiques, *Discrete Math. 25*, 1–5.

Bean, D. R., A. Ehrenfeucht, and G. F. McNulty, 1979, Avoidable patterns in strings of symbols, *Pacific J. Math. 85*, 261–294.

Behrend, F., 1946, On sets of integers which contain no three terms in arithmetical progression, *Proc. Nat. Acad. Sci. U.S.A. 32*, 331–332.

Behrend, F., 1938, On sequences of integers containing no arithmetic progression, *Časopis Pěst. Mat. 67*, 235–239.

Berlekamp, E. R., 1968, A construction for partitions which avoid long arithmetic progressions, *Canad. Math. Bull. 11*, 409–414.

Berstel, J., 1979, Sur les Mots sans carrés définis par morphisme, in H. Maurer (ed.), Automata, Languages, and Programming, 6th Coll., Lecture Notes Computer Science, vol. 71, Springer–Verlag, Berlin, pp. 16–25.

Berstel, J., D. Perrin, J. F. Perrot, and A. Restivo, 1979, Sur le Théorème du défaut, *J. Algebra 60*, 169–180.

Bourbaki, N., 1971, *Groupes et Algèbres de Lie*, Hermann, vol. 1, 1971, chaps. 1 and 2, also vol. 2.

Braunholtz, C., 1963, An infinite sequence of three symbols with no adjacents repeats, *Am. Math. Monthly 70*, 675–676.

Brown, T. C., 1969, On van der Waerden's theorem on arithmetic progression, *Notices Am. Math Soc. 16*, 245.

Brown, T. C., 1969, On van der Waerden's theorem on arithmetic progression, *Notices Am. Math Soc. 16*, 245.

Brown, T. C., 1971, An interesting combinatorial method in the theory of locally finite semigroups, *Pacific J. of Math. 36*, 285–289.

Brown, T. C., 1971, Is there a sequence on four symbols in which no two adjacent segments are permutations of one another? *Am. Math. Monthly*, *78*, 886–888.

Brown, T. C., 1981, On van der Waerden's theorem and the theorem of Paris and Harrington, *J. Comb. Theory, Ser. A, 30*, 108–111.

Brzozowski, J. A., K. Culik, II, and A. Gabrielan, 1971, Classification of noncouting events, *J. Comput. System Sci.*, *5*, 41–53.

Cartier, P., and D. Foata, 1969, *Problèmes combinatoires de commutation et réarrangements*, Lecture Notes in Math., vol. 85, Springer–Verlag, Berlin.

Catalan, E., 1838, Note sur une equation aux différences finies, *J. Math. Pures Appl.* 508–516.

Cesari, Y., and M. Vincent, 1978, Une Caractérisation des mots périodiques, *C.R. Acad. Sci. Paris*, *286*, A, 1175–1177.

Chen, K. T., 1957, Integration of paths, geometric invariants, and a generalized Baker–Hausdorff formula, *Ann. Math.*, *65*, 163–178.

Chen, K. T., R. H. Fox, and R. C. Lyndon, 1958, Free differential calculus, IV — The quotient groups of the lower central series, *Ann. Math.*, *68*, 81–95.

Chottin, L., 1975, Une Preuve combinatoire de la formule de Lagrange à deux variables, *Discrete Math.*, *13*, 214–224.

Christol, C., T. Kamae, M. Mendès-France, and G. Rauzy, 1980, Suites algébriques, automates et substitutions, *Bull. Soc. Math. France*, *108*, 401–419.

Cohn, P. M., 1962, On subsemigroups of free semigroups, *Proc. Am. Math. Soc.*, *13*, 347–351.

Cohn, P. M., 1971, *Free Rings and Their Relations*, Academic Press, New York.

Comtet, L., 1970, *Analyse combinatoire*, vols. 1 and 2, Presses Universitaires de France, Paris.

Content, M., F. Lemay, and P. Leroux, 1980, Catégories de Möbius et fonctorialitiés: un cadre général pour l'inversion de Möbius, *J. Comb. Theory*, *Ser. A*, *28*, 169–190.

Conway, J. H., 1971, *Regular Algebra and Finite Machines*, Chapman and Hall.

Crochemore, M., 1981, Optimal determination of repetitions in a string, *Inf. Proc. Letters*, *12*, 244–249.

Crochemore, M., 1982, A sharp characterization of square free morphisms, *Theoret. Comput. Sci.*, to appear.

Culik, K., and J. Karhumäki, 1980, On the equality sets for homomorphisms on free monoids with two generators, *RAIRO Informat. Théor.*, *14*, 349–369.

Culik, K. and A. Salomaa, 1980, Test sets and checking words for homomorphism equivalence, *J. Comput. System Sci.*, *20*, 379–395.

Dean, R., 1965, A sequence without repeats on x, x^{-1}, y, y^{-1}, *Am. Math. Monthly*, *72*, 383–385.

Dejean, F., 1972, Sur un Théorème de Thue, *J. Comb. Theory*, *Ser. A*, *13*, 90–99.

Dekking, F. M., 1976, On repetition of blocks in binary sequences, *J. Comb. Theory*, *Ser. A*, *20*, 292–299.

Dekking, F. M., 1978, The spectrum of dynamical systems arising from substitutions of constant length, *Z. Wahrscheinlichkeitstheorie und Verw. Gebiete*, *41*, 221–239.

Dekking, F. M., 1979, Strongly non-repetitive sequences and progression-free sets, *J. Comb. Theory*, *Ser. A*, *27*, 181–185.

Deuber, W., 1982, On van der Waerden's theorem on arithmetic progressions, *J. Comb. Theory*, *Ser. A*, *32*, 115–118.

De Luca, A., 1979, Rearrangements of words and equations in free monoids, in Proceedings of the Colloquium *Codages et Transductions*, Florence.

Dickson, L. E., 1903, Finiteness of the odd perfect and primitive abundant numbers with r distinct prime factors, *Am. J. Math.*, *35*, 413–422.

Dumont, D., 1981, Une approche combinatoire des fonctions elliptiques de Jacobi, *Advances in Math.*, *41*, 1–39.

Dumont, D., and G. Viennot, 1980, A combinatorial interpretation of the Seidel generation of Genocchi numbers, *Ann. Discrete Math.*, *6*, 77–87.

Duval, J. P., 1979a, Périodes et répétitions des mots du monoïde libre, *Theoret. Comput. Sci.*, *9*, 17–26.

Duval, J. P., 1979b, Une Caractérisation des fonctions périodiques, *C.R. Acad. Sci. Paris*, *289*, 185–187.

Duval, J. P., 1980, Mots de Lyndon et périodicité, *RAIRO Informat. Théor.*, *14*, 181–191.

Duval, J. P., 1980, *Contribution à la combinatoire du monoïde libre*, thesis, University of Rouen.

Duval, J. P., 1981, A remark on the Knuth–Morris–Pratt string searching algorithm, unpublished manuscript.

Ehrenfeucht, A., J. Karhumäki and G. Rozenberg, 1982, On binary equality sets and a solution to the Ehrenfeucht conjecture in the binary case, unpublished manuscript.

Ehrenfeucht, A., and G. Rozenberg, 1978, Elementary homomorphisms and a solution of the DOL sequence equivalence problem, *Theoret. Comput. Sci.*, *7*, 169–183.

Ehrenfeucht, A., and G. Rozenberg, 1981, On the subword complexity of square-free DOL languages, *Theoret. Comp. Sci.*, *16*, 25–32.

Ehrenfeucht, A., and G. Rozenberg, 1981, On the separating power of EOTOL, *RAIRO Informat. Théor.*, to appear.

Ehrenfeucht, A., and D. M. Silberger, 1979, Periodicity and unbordered words, *Discrete Math.*, *26*, 101–109.

Eilenberg, S., 1974, *Automata, Languages, and Machines*, vol. A, Academic Press, New York

Eilenberg, S., 1976, *Automata, Languages, and Machines*, vol. B, Academic Press, New York

Entringer, R., D. Jackson, and J. Schatz, 1974, On nonrepetitive sequences, *J. Comb. Theory*, *Ser. A*, *16*, 159–164.

Erdös, P., 1963, Quelques Problèmes de la théorie des nombres, *Monographie de l'Enseignement Math.*, *6*, 81–135.

Erdös, P., and P. Turan, 1936, On some sequences of integers, *J. London Math. Soc. 11*, 261–264.

Erdös, P., 1965, Some recent advances and current problems in number theory, *On Modern Mathematics*, vol. 3, J. Wiley and Sons, New York, pp. 196, 226.

Erdös, P., 1977, Problems and results on combinatorial number theory III, in M. B. Nathanson (ed.), *Number Theory Day*, Lecture Notes in Mathematics, vol. 626 Springer–Verlag, Berlin, pp. 43–72.

Erdös, P., and J. Spencer, 1974, *Probabilistic Methods in Combinatorics*, Academic Press, New York, p. 31.

Erdös, P., and R. L. Graham, 1979, Old and new problems and results in combinatorial number theory: van der Waerden's theorem and related topics, *L'enseignement mathématique*, *2nd series*, *25*, 3–4.

Evdokimov, A. A., 1968, Strongly asymmetric sequences generated by a finite number of Symbols, *Dokl. Akad. Nauk SSSR*, *179*, 1268–1271 (also *Soviet Math. Dokl. 9* (1968), 536–539).

Farrel, R. H., Notes on a combinatorial theorem of Bohnenblust, *Duke Math. J.*, *32*, 333–339.

Fine, N. J., and H. S. Wilf, 1965, Uniqueness theorem for periodic functions, *Proc. Am. Math. Soc.*, *16*, 109–114.

Fliess, M., 1981, Fonctionnelles causales non-linéaires et indéterminées non-commutative, *Bull. Soc. Math. France*, *109*, 3–40

Foata, D., 1965, Etude algébrique de certains problèmes d'analyse combinatoire et du calcul des probabilités, *Publ. Inst. Statist. Univ. Paris*, *14*, 81–241.

Foata, D., 1968, On the Netto inversion number of a sequence, *Proc. Am. Math. Soc.*, *19*, 236–240.

Foata, D., 1977, Distributions Eulériennes et Mahoniennes sur le groupe des permutations, in M. Aigner (ed.), *Higher Combinatorics*, 27–49, D. Reidel, Boston, Berlin Combinatorics Symposium, 1976.

Foata, D., 1979, A non-commutative version of the matrix inversion formula, *Advances in Math.*, *31*, 330–349.

Foata, D., 1980, A combinatorial proof of Jacobi's identity, *Ann. Discrete Math.*, *6*, 125–135.

Foata, D., and M. P. Schützenberger, 1970, *Théorie géométrique des polynômes eulériens*, Lecture Notes in Math., vol. 138, Springer–Verlag, Berlin.

Foata, D., and M. P. Schützenberger, 1971, On the principle of equivalence of Sparre-Andersen, *Math. Scand.*, *28*, 308–316.

Foata, D., and M. P. Schützenberger, 1978, Major index and inversion number of permutations, *Math. Nachr.*, *83*, 143–159.

Fox, R. H., 1953, Free differential calculus, *Ann. Math.*, *57*, 547–560.

Fürstenberg, H., 1977, Ergodic behavior of diagonal measures and a theorem of Szemerédi on arithmetic progressions, *J. d'Analyse Math.*, *31*, 204–256.

Fürstenberg, H., and Y. Katznelson, 1978, An ergodic Szemerédi theorem for commuting transformations, *J. Analyse Math.*, *34*, 275–291.

Fürstenberg, H., and B. Weiss, 1978, Topological dynamics and combinatorial number theory, *J. Analyse Math.*, *34*, 61–85.

Gardelle, J., and G. Th. Guilbaud, 1964, Cadences, *Math. Sci. Humaines*, *9*, 31–38.

Garsia, A. M., and I. M. Gessel, 1979, Permutation statistics and partitions, *Advances in Math.*, *31*, 288–305.

Gerver, J. L., 1977, The sum of the reciprocals of a set of integers with no arithmetic progression of k terms, *Proc. Am. Math. Soc.*, *62*, 211–214.

Gessel, I. M., 1977, *Generating functions and enumeration of sequences*, Ph.D. thesis, Department of Math., Massachusetts Institute of Technology, Cambridge, Mass., 111 pp.

Girard, J. Y., 1982, *Proof Theory and Logical Complexity*, Bibliopolis, Naples, to appear.

Good, I. J., 1960, Generalization to several variables of Lagrange's expansion, *Math. Proc. Camb. Phil. Soc.*, *56*, 367–380.

Gottschalk, W., and G. Hedlund, 1955, *Topological Dynamics*, Am. Math. Soc. Colloq. Publ., vol. 36.

Gottschalk, W., and G. Hedlund, 1964, A characterization of the Morse minimal set, *Proc. Amer. Math. Soc.*, *15*, 70–74.

Gouyou-Beauchamps, D., 1975, Deux propriétés combinatoires du langage de Lukasiewicz, *RAIRO Informat. Théor.*, *9*, 13–24.

Graham, R. L., and B. L. Rothschild, 1974, A short proof of van der Waerden's theorem on arithmetic progressions, *Proc. Amer. Math. Soc.*, *42*, 385–386.

Graham, R. L., B. L. Rotschild, and J. H. Spencer, 1980, *Ramsey Theory*, J. Wiley and Sons, New York.

Green, J. A., and D. Rees, 1952, On semigroups in which $x^r = x$, *Mat. Proc. Cambridge Phil. Soc.*, *48*, 35–40.

Haines, L. H., 1969, On free monoids partially ordered by embedding, *J. Comb. Theory*, *6*, 94–98.

Hall, M., 1959, *Theory of Groups*, MacMillan, New York.

Harary, F., Prins, and W. Tutte, 1964, The number of plane trees, *Indag. Math.*, *26*, 319–329.

Hawkins, D., and W. Mientka, 1956, On sequences which contain no repetition, *Math. Student*, *24*, 185–187.

Hedlund, G., 1967, Remarks on the work of Axel Thue on sequences, *Nordisk Mat. Tidskr.*, *15*, 147–150.

Herstein, I. N., 1968, *Non commutative rings*, Carus Mathematical Monograph, John Wiley and Sons, New York.

Higman, G., 1952, Ordering by divisibility in abstract algebras, *Proc. London Math. Soc.*, *2*, 326–336.

Hmelevskii, Y. I., 1976, Equations in free semigroups, *Proc. Steklov and Inst. Mat.*, Am. Math. Soc. Transl., *107*, 272.

Istrail, S., 1977, On irreductible languages and non rational numbers, *Bull. Mat. Soc. Sci. Mat. R.S. Roumanie*, *21*, 301–308.

Jacob, G., 1980, Un Théorème de factorisation des produits d'endomorphismes, *J. Algebra*, *63*, 389–412.

Jacobson, N., 1962, *Lie Algebras*, Interscience, New York.

Jullien, P., 1968, Sur un Théorème d'extension dans la théorie des mots, *C.R. Acad. Sci. Paris*, *A.*, *266*, 851–854.

Justin, J., 1969, Propriétés combinatoires de certains semigroupes, *C.R. Acad. Sci. Paris, A*, *269*, 1113–1115.

Justin, J., 1970, Semigroupes á générations bornées, in *Problimes Mathématiques de la Théorie des Automates*, Séminaire Schützenberger, Lentin, Nivat 69170, Institut Henri Poincaré, Paris, exposé No 7, 10 p.

Justin, J., 1971a, Sur une Construction de Bruck et Reilly, *Semigroup Forum*, *3*, 148–155.

Justin, J., 1971b, Groupes et semigroupes à croissance linéaire, *C.R. Acad. Sci. Paris, A*, *273*, 212–214.

Justin, J., 1971c, Semigroupes répétitifs, in *Logique et Automates*, Séminaires I. R. I. A., Institut de Recherche d'Informatique et d'Automatique, Le Chesnay, France, 101–108.

Justin, J., 1972a, Généralisation du théorème de van der Waerden sur les semigroups répétitifs, *J. Comb. Theory*, *12*, 357–367.

Justin, J., 1972b, Characterization of the repetitive commutative semigroups, *J. Algebra*, *21*, 87–90.

Justin, J., 1981, Groupes linéaires répétitifs, *C.R. Acad. Sci. Paris, A*, to appear.

Kaplansky, I., 1969, *Fields and Rings*, Chicago Lectures in Mathematics, Chicago, London.

Khinchin, A. Y., 1952, *Three pearls of number theory*, Graylock Press, Rochester, N.Y.

Knuth, D. E., 1968, *The Art of Computer Programming*, *vol. 1: Fundamental Algorithms*, Addison-Wesley, Reading, Mass.

Knuth, D. E., J. H. Morris, and V. R. Pratt, 1977, Fast pattern matching in strings, *SIAM J. Comput.*, *6*, 323–350.

Kruskal, J. B., 1972, The theory of well quasi-ordering: a frequently discovered concept, *J. of Comb. Theory, Ser. A*, *13*, 297–305.

Lallement, G., 1979, *Semigroups and Combinatorial Applications*, J. Wiley and Sons, New York.

Leech, J., 1957, A problem on strings of beads, note 2726, *Math. Gaz.*, *41*, 277–278.

Lentin, A., 1972a, *Equations dans les Monoïdes Libres*, Gauthier-Villars, Paris.

Lentin, A., 1972b, Equations in free monoids, in M. Nivat (ed.), *Automata Languages and Programming*, North Holland, Amsterdam, pp. 67–85.

Lentin, A., and M. P. Schützenberger, 1967, A combinatorial problem in the theory of free monoids, in R. C. Bose and T. E. Dowling (eds.), *Combinarorial Mathematics*, North Carolina Press, Chapel Hill, N.C., pp. 112–144.

Li, S. Y. R., 1976, Annihilators in nonrepetitive semigroups, *Studies in Appl. Math.*, *55*, 83–85.

Lyndon, R. C., 1954, On Burnside problem I, *Trans. Am. Math. Soc.*, *77*, 202–215.

Lyndon, R. C., 1955, On Burnside problem II, *Trans. Am. Math. Soc.*, *78*, 329–332.

Lyndon, R. C., 1959, The equation $a^2b^2 = c^2$ in free groups, *Michigan Math. J.*, *6*, 155–164.

Lyndon, R. C., 1960, Equations in free groups, *Trans. Am. Math. Soc.*, *96*, 445–457.

Lyndon, R. C., and P. E. Schupp, 1977, *Combinatorial Group Theory*, Springer–Verlag, Berlin.

Lyndon, R. C., and M. P. Schützenberger, 1962, The equation $a^m = b^n c^p$ in a free group, *Michigan Math. J.*, *9*, 289–298.

MacMahon, P. A., 1913, The indices of permutations and the derivation therefrom of functions of a single variable associated with the permutations of any assemblage of objects, *Am. J. Math.*, *35*, 281–322.

MacMahon, P. A., 1915, *Combinatory Analysis*, vol. 1, Cambridge Univ. Press, Cambridge, England.

MacMahon, P. A., 1916, Two applications of general theorems in combinatory analysis, *Proc. London Math. Soc.*, *15*, 314–321.

Magnus, W., A. Karrass, and D. Solitar, 1976, *Combinatorial Group Theory*, 2nd ed., Dover, New York.

Main, M., and R. Lorentz, 1979, An $O(n \log m)$ algorithm for finding repetition in a string, Tech. Rept. CS-79-056, Computer Science, Washington State University.

Makanin, G. S., 1976, On the rank of equations in four unknowns in a free semigroup, *Mat. Sb.*, *100*, 285–311 (English trans. in *Math. USSR Sb.*, *29*, 257–280).

Makanin, G. S., 1977, The problem of solvability of equation in a free semigroup, *Mat. Sb.*, *103*, 147–236 (English trans. in *Math. USSR Sb.*, *32*, 129–198).

Makowski, A., 1959, Remark on a paper of Erdös and Turan, *J. London Math. Soc.*, *34*, 480.

McNaughton, R., and I. Zalcstein, 1975, The Burnside problem for semigroups, *J. Algebra*, *34*, 292–299.

Meier-Wunderli, H., 1952, Note on a basis of P. Hall for the higher commutators in free groups, *Comm. Math. Helv.*, *26*, 1–5.

Michel, J., 1974, Bases des algèbres de Lie libres et série de Hausdorff, *Séminaire Dubreil* (*algèbre*), *27, no. 6*, 9 p., Institut Henri Poincaré, Paris.

Morse, M., 1921, Recurrent geodesics on a surface of negative curvature, *Trans. Am. Math. Soc.*, *22*, 84–100.

Morse, M., 1938, A solution of the problem of infinite play in chess, *Bull. Am. Math. Soc.*, *44*, 632.

Morse, M., and G. Hedlund, 1938, Symbolic dynamics, *Am. J. Math.*, *60*, 815–866.

Morse, M., and G. Hedlund, 1944, Unending chess, symbolic dynamics and a problem in semigroups, *Duke Math. J.*, *11*, 1–7.

Moser, L., 1953, On non averaging sets of integers, *Canad. J. Math.*, *5*, 245–252.

Moser, L., 1960, On a theorem of van der Waerden, *Canad. Math. Bull.*, *3*, 23–25.

Moser, L., 1970, Problem 170, *Canad. Math. Bull.*, *13*, 268.

Nash Williams, C., 1963, On well quasi-ordering finite trees, *Proc. Cambridge Phil. Soc.*, *59*, 833–835.

Narayana, 1959, A partial order and its applications to probability, *Sankhyà*, *21*, 91–98.

Ochsenschläger, P., 1981a, Binomialkoeffizenten und Shuffle-Zahlen, Technischer Bericht, Fachbereich Informatik, T. H. Darmstadt.

Ochsenschläger, P., 1981b, Eine Klasse von Thue-Folgen, Technischer Bericht, Fachbereich Informatik, T. H. Darmstadt.

Pansiot, J. J., 1981, The Morse sequence and iterated morphisms, *Information Proc. Letters*, *12*, 68–70.

Pecuchet, J. P., 1981, Sur la Détermination du rang d'une équation dans le monoide libre, *Theoret. Comput. Sci.*, *16*, 337–340.

Perrin, D. and G. Viennot, 1982, A note on shuffle algebras, unpublished manuscript.

Piollet, D., 1975, Equations quadratiques dans le groupe libre, *J. Algebra*, *33*, 395–404.

Pirillo, G., 1981, Thèse de 3ème cycle, Université Paris 7.

Pleasants, P. A., 1970, Non-repetitive sequences, *Mat. Proc. Cambridge, Phil. Soc. 68*, 267–274.

Procesi, C., 1973, *Rings with Polynomial Identities*, Dekker, New York.

Putcha, M. S., 1979, Generalization of Lentin's theory of principal solutions of word equations in free semigroups to free product of copies of positive reals under addition, *Pacific J. Maths.*, *83*, 253–268.

Rabung, J. B., 1970, A note on van der Waerden's theorem on arithmetic progressions, *Math. Research Center Report* 70-6; Math and Information Sciences Division, *Naval Research Laboratory Report* 70.73, Washington D.C.

Ramsey, F. P., 1930, On a problem of formal logic, *Proc. London Math. Soc.*, *2nd ser., 30*, 264–286.

Raney, G. N., 1960, Functional composition patterns and power series reversion, *Trans. Am. Math. Soc.*, *94*, 441–451.

Rankin, R. A., Sets of integers containing not more than a given number of terms in arithmetical progression, *Proc. Roy. Soc. Edinburg, Sect. A, 65*, 332–344.

Rawlings, D., 1981, Permutation and multipermutation statistics, *European J. Combinatorics 2*, 67–78.

Ree, R., 1958, Lie elements and an algebra associated with shuffles, *Ann. Math.*, *68*, 210–220.

Reutenauer, C., 1980, An Ogden-like iteration lemma for rational power series, *Acta Informat.*, *13*, 189–197.

Riordan, J., 1958, *An Introduction to Combinatorial Analysis*, John Wiley, New York.

Riordan, J., 1968, *Combinatorial Identities*, John Wiley, New York.

Rota, G. C., 1964, On the foundations of combinatorial theory: I. Theory of Möbius functions, *Z. Warscheinlichkeits theorie*, *2*, 340–368.

Rota, G. C., 1978, *Studies in Combinatorics*, Studies in Mathematics, vol. 17, Academic Press, New York.

Roth, K. F., 1952, Sur quelques ensembles d'entiers, *C.R. Acad. Sci. Paris, A, 234*, 388–390.

Roth, K. F., 1970, Irregularities of sequences relative to arithmetic progressions, III, *J. Number Theory*, *2*, 125–142.

Roth, K. F., 1972, Irregularities of sequences relative to arithmetic progression, IV, *Period. Math Hungar.*, *2*, 301–326.

Rowen, L. H., 1980, *Polynomial Identities in Ring Theory*, Academic Press, New York.

Rozenberg, G., and A. Salomaa, 1980, *The Mathematical Theory of L Systems*, Academic Press, New York.

Salem, R., and D. C. Spencer, 1942, On sets of integers which contain no three terms in arithmetical progression, *Proc. Nat. Acad. Sci. U.S.A.*, *28*, 561–563.

Schmidt, W. M., 1962, Two combinatorial theorems on arithmetic progressions, *Duke Math. J.*, *29*, 129–140.

Schützenberger, M. P., 1956, Une Théorie algébrique du codage, *Séminaire Dubreil–Pisot*, année 55–56, Institut Henri Poincaré, Paris.

Schützenberger, M. P., 1958, Sur une Propriété combinatoire des algèbres de Lie libres pouvant être utilisés dans un problème de Mathématiques Appliquées, *Séminaire Dubreil–Pisot*, année 58–59, Inst. Henri Poincaré, Paris.

Schützenberger, M. P., 1964, On the synchronizing properties of certain prefix codes, *Information and Control*, *7*, 23–36.

Schützenberger, M. P., 1965, On a factorization of free monoids, *Proc. Am. Math. Soc.*, *16*, 21–24.

Schützenberger, M. P., 1971, Le théorème de Lagrange selon Raney, in *Logiques et Automates*, Publications Inst. Rech. Informatique et Automatique, Rocquencourt, France.

Shapiro, H. N., and G. H. Sparer, 1972, Composite values of exponential and related sequences, *Comm. Pure Appl. Math.*, *25*, 569–615.

Shepherdson, J. C., 1958, Note 2813: A historical note on Note 2726, *Math. Gaz.*, *42*, 306.

Shirshov, A. I., 1957a, On certain non associative nil rings and algebraic algebras, *Mat. Sb.*, *41*, 381–394.

Shirshov, A. I., 1957b, On rings with identity relations, *Mat. Sb.*, *43*, 277–283.

Shirshov, A. I., 1958, On free Lie rings, *Math. Sb.*, *45*, 113–122.

Simon, I., 1972, *Hierarchies of events with dot-depth one*, Ph.D. Thesis, University of Waterloo, Canada.

Simon, I., 1975, Piecewise testable events, in H. Brakhage (ed.), *Automata Theory and Formal Languages*, Lecture Notes in Computer Science, vol. 33, Springer–Verlag, Berlin, pp. 214–222.

Simon, I., 1980, Conditions de finitude pour des semigroupes, *C. R. Acad. Sci. Paris*, *290, A*, 1081–82.

Sparre-Andersen, E., 1962, The equivalence principle in the theory of sums of random variables, *Colloquium Aarhus*, August 1–10, 1962, 13–16.

Spehner, J. C., 1976, *Quelques Problèmes d'extension, de conjugaison et de présentation des sous-monoïdes d'un monoïde libre*, Thèse de Doctorat d'Etat, Paris.

Spehner, J. C., 1981, Les présentations des sous-monoïdes de rang 3 d'un monoïde libre, in H. Jürgensen, H. Y. Weinert, and M. Petrich (eds.), *Semigroups*, Lecture Notes in Math., *855*, Springer–Verlag, Berlin, 116–155.

Spitzer, F., 1956, A combinatorial lemma and its applications to probability theory, *Trans. Am. Math. Soc.*, *82*, 441–452.

Stanley, R. P., 1976, Binomial posets, Möbius inversion and permutation enumeration, *J. Comb. Theory, Ser. A*, *20*, 336–356.

Strehl, V., 1981, Une q-extension d'une congruence de Lucas, unpublished manuscript.

Szemerédi, E., 1969, On sets of integers containing no four elements in arithmetic progression, *Acta Math. Acad. Sci. Hungar*, *20*, 89–104.

Szemerédi, E., 1975, On sets of integers containing no k elements in arithmetic progression, *Acta Arith.*, *27*, 199–245.

Thouvenot, J. P., 1977, La démonstration de Fürstenberg du théorème de Szemerédi sur les progressions arithmétiques, *Séminaire Bourbaki*, *30*, no. 518 (1977/78), 11 pp.

Thue, A., 1906, Über unendliche Zeichenreihen, *Norske Vid. Selsk. Skr. I. Mat. Nat. Kl.*, Christiania no. 7, 1–22.

Thue, A., 1912, Über die gegenseitige Lage gleicher Teile gewisser Zeichenreihen, *Norske Vid. Selsk. Skr. I. Mat-Nat. Kl.*, Christiania no. 1, 1–67.

van der Waerden, B. L., 1927, Beweis einer Baudet'schen Vermutung, *Nieuw Arch. Wisk.*, *15*, 212–216.

van der Waerden, B. L., 1965, Wie der Beweis der Vermutung von Baudet gefunden wurde, *Abhandlungen des Mathematischen Seminars der Hanseatischen Universität Hamburg*, pp. 6–15; also published as: [How the proof of Baudet's conjecture was found, *Studies in Pure Mathematics*, Academic Press, New York, 1971, pp. 251–260.]

Viennot, G., 1974, *Algèbres de Lie libres et monoïdes libres*, thèse, Université Paris VII.

Viennot, G., 1978, Algèbres de Lie libres et monoïdes libres, *Lecture Notes in Math.*, *691*, Springer–Verlag, Berlin.

Viennot, G., 1978, Maximal chains of subwords and up–down sequences of permutations, unpublished manuscript.

Wagstaff, S. S., 1979, Some questions about arithmetic progressions. *Am. Math. Monthly*, *86*, 579–582.

Weintraub, S., 1977, Seventeen primes in arithmetic progression, *Math. Comp.*, *31*, 1030.

Witt, E., 1951, Ein kombinatorischer Satz der Elementargeometric, *Math. Nachr.*, B. 1–6, H. 5, 261–262.

Yaglom, A. M., and I. M. Yaglom, 1967, *Challenging Mathematical Problems*, vol. 2, Holden Day, San Francisco, pp. 12, 94–97, 203–204.

Zech, Th., 1958, Wiederholungsfreie Folgen, *Z. Angew. Math. Mech.*, *38*, 206–209.

Index

Index

DEMCO